Airborne Pollutants in Museums, Galleries, and Archives:
Risk Assessment, Control Strategies, and Preservation Management

BY JEAN TÉTREAULT

 Canadian   Patrimoine
Heritage   canadien

Available from:

Canadian Conservation Institute
1030 Innes Road
Ottawa ON K1A 0M5
Canada

NATIONAL LIBRARY OF CANADA CATALOGUING IN PUBLICATION

Tétreault, Jean, 1963-
Airborne pollutants in museums, galleries and archives : risk assessment, control strategies and preservation management / by Jean Tétreault.

Issued also in French under title: Polluants dans les musées et les archives, évaluation des risques, stratégies de contrôle et gestion de la préservation.
Includes bibliographical references.
ISBN 0-662-34059-0
Cat. no. NM95-59/9-2003E

1. Museum conservation methods. 2. Indoor air pollution. 3. Air quality management. I. Canadian Conservation Institute II. Title.

AM145.T47 2003   069'.53   C2003-902721-X

DISCLAIMER

The information provided in this book is based on current scientific and technical understanding of the issues presented. Following the advice given in this book will not necessarily provide complete protection in all situations or against all adverse effects that may be caused by airborne pollutants.

COVER

A Fantastic Interior by Giovanni Battista Piranesi (c.1750 — 1759).
Pen and brown ink wash brown over red chalk on laid paper.
National Galley of Canada, Ottawa.

Airborne Pollutants in Museums, Galleries, and Archives:
Risk Assessment, Control Strategies, and Preservation Management

BY JEAN TÉTREAULT

Acknowledgements

The author sincerely thanks the following individuals for reviewing the draft document
and submitting valuable comments:

- Carole Dignard, Cliff McCawley, Stefan Michalski, Lyndsie Selwyn, and Scott Williams (CCI)
- Jonathan Ashley-Smith (Victoria and Albert Museum, London, United Kingdom)
- Agnes Brokerhof and Frank Ligterink (Netherlands Institute for Cultural Heritage, Amsterdam, The Netherlands)
- Andrew Calver (Museum of London, London, United Kingdom)
- William Esposito (Ambient Group, Inc., Glen Cove, New York, United States)
- Jost Kames (McLeod Russel Filter AG, Uster, Switzerland)
- Gemma Kerr (InAir Environmental Ltd., Ottawa, Canada)
- Bertrand Lavédrine (Centre de recherches sur la conservation des documents graphiques, Paris, France)
- Morten Ryhl-Svendsen (The National Museum of Denmark, Brede, Denmark)
- Rob Waller (Canadian Museum of Nature, Ottawa, Canada)
- Steven Weinstraub (Art Preservation Services, New York, United States)

The author also acknowledges Dianne Perrier and Barbara Patterson for the editing the English text;
Françoise Guyot of Services T&A Inc. for translating the English text to French; Linda Leclerc for editing
the French text; Jeremy Powell for taking some of the photographs; Joe Iraci, Paul Bégin, and Fiona Jones
for providing examples of damaged objects; Scott Williams for providing examples of damaged objects and
identifying some cellulose compounds; Bill Conly for re-drawing some of the illustrations; Carl Bigras
for scanning and editing the photographs; Sophie Georgiev for designing and laying out the publication;
and Lucie Forgues for ordering reference documents so promptly.

TABLE OF CONTENTS

Note: Information supplementing this book can be found at
<www.cci-icc.gc.ca/links/pollutants/index_e.shtml>

PREFACE

Over the last few decades, many observations of the impact of airborne pollutants on materials in museums and laboratories have provided a better understanding of potential interactions and their kinetics. At the same time, the risk management field has developed new approaches for evaluating risk. The combination of these two sciences has provided valuable tools for preservation assessment and for the development of pragmatic strategies and policies to counter pollution in the context of museums, galleries, and archives. These tools can be used by museum directors, building and collection managers, conservation professionals, material scientists, exhibit designers, HVAC engineers, indoor air consultants, and architects to help them make decisions regarding the preservation of collections.

The goals of this book are to:
- define key airborne pollutants for indoor museum environments
- supply tools based on the exposure–effect relationship to assess the risk to collections exposed to pollutants
- establish guidelines for control strategies that give flexible, pragmatic solutions
- offer guidelines for assessing the degree of protection of collections
- provide a basic tool for cost–benefit analysis that fulfils the principles and the policy of the museum
- propose an investigative approach for evaluating the degree of preservation of collections in terms of the aggressiveness of pollutants and for identifying causes of damage

Even though the effect of pollutants on some materials remains equivocal and the effectiveness of some control strategies is not fully known, this book provides a basis for making judgments about a situation and better cost–benefit decisions.

This book does not cover the effects of pollutants on human health and on outdoor cultural property, the effects of non-airborne pollutants (such as pollutants transferred from one material to another at a contact point), or the degradation of objects through their own internal pollutants.

Jean Tétreault

LIST OF BOXES

LIST OF TABLES

LIST OF FIGURES

* Pages numbers in **bold** indicate colour photographs.

ACRONYMS AND ABBREVIATIONS

ACS	American Chemical Society
AE/h	Air exchange per hour
Ag	Silver
AHP	Analytical hierarchy process
AIC	American Institute for Conservation of Historic and Artistic Works
AINC	Australian Network for Information on Cellulose Acetate
Al	Aluminum
APT	Association for Preservation Technology International
ASHRAE	American Society of Heating, Refrigeration and Air-Conditioning Engineers
ASTM	American Society for Testing and Materials
Avg.	Average
CAC	Canadian Association for Conservation of Cultural Property
$CaCO_3$	Calcium carbonate
CAPC	Canadian Association of Professional Conservators
CCI	Canadian Conservation Institute
CEN	European Committee for Standardisation
CH_2O	Formaldehyde
CH_3COOH	Acetic acid
CIE	International Commission on Illumination (Commission Internationale d'Éclairage)
Cl_2	Chlorine
CO_2	Carbon dioxide
Cu	Copper
DEAE	Diethylamino ethanol
EC	Environment Canada
EEA	European Environment Agency
EPA	Environmental Protection Agency, United States
GC/FID	Gas chromatography coupled to a flame ionization detector
GC/MS	Gas chromatography coupled to a mass spectrometer
h	Hour
H_2O	Water
H_2S	Hydrogen sulphide
H_2SO_4	Sulphuric acid
HC	Health Canada
HCOOH	Formic acid
HEPA	High-efficiency particulate air
HNO_3	Nitric acid
HPLC	High-performance liquid chromatography
HVAC system	Heating, ventilating, and air-conditioning system
I/O	Indoor/outdoor ratio
ISA	Instrument Society of America
ISO	International Organization for Standardization
KCO_3	Potassium carbonate
kg	Kilogram
$KMnO_4$	Potassium permanganate
KOH	Potassium hydroxide
LOAED	Lowest observed adverse effect dose
LOAEL	Lowest observed adverse effect level
MERV	Minimum efficiency reporting value
MOE	Ministry of the Environment, Ontario
MSDS	Material safety data sheet
N	Air exchange rate
N_2O	Nitrous oxide
$NaHCO_3$	Sodium bicarbonate
nd	No date
NFPA	National Fire Protection Association
NH_3	Ammonia
nm	Nanometre
NO	Nitric oxide
NO_2	Nitrogen dioxide
NO_X	Nitrogen oxides
NOAEL	No observed adverse effect level
NV	Natural ventilation
O_2	Oxygen
O_3	Ozone
OCS	Carbonyl sulphide
ODA	Octadecylamine
PAN	Peroxyacetyl nitrate
Pb	Lead
PLV	Pump large (air) volume
$PM_{2.5}$	Suspended particle matter of 2.5-μm diameter or less
PM_{10}	Suspended particle matter of 10-μm diameter or less
ppb or ppbv	Particles per billion (volume/volume)
ppm or ppmv	Particles per million (volume/volume)
PVAc	Poly(vinyl acetate)
PVC	Poly(vinyl chloride)
RH	Relative humidity
RTV	Room temperature vulcanization
SO_2	Sulphur dioxide
SO_X	Sulphur oxides
SPME	Solid-phase microextraction
Std	Standard deviation
TVOC	Total volatile organic compounds
UK-IIC	United Kingdom, International Institute of Conservation of Historic and Artistic Works
UV	Ultraviolet
VAV	Variable air volume
VOC	Volatile organic compound
WRI	World Resources Institute
yr	Year
$\mu g\ m^{-3}$	Micrograms per cubic metre
$\mu g\ m^{-3}\ yr$	Micrograms per cubic metre multiplied by year (dose)
μm	Micrometre

KEY AIRBORNE POLLUTANTS 1

Many airborne pollutants[*1] cause adverse effects[*] on collections in an indoor environment. These pollutants have been arranged into eight distinct chemical groups; particles (dust) form a ninth group. Table 1 shows the sources and effects of these different groups of pollutants on objects. Pollutants originate both from outside and inside a building. Indoors, pollutants are typically from products, indoor activities (such as cleaning), visitors, and even objects in the collection. Outdoors, pollutants are mainly related to human activities such as industrial processes and vehicular traffic. Table 2 shows the effects of pollutants on objects in an indoor environment or in model experimental conditions. Different objects are susceptible to different airborne pollutants and deteriorate at varying rates depending on the parameters involved. In some cases, damage is caused by more than one pollutant. For example, the analysis of corrosion compounds on a metal often confirms the presence of more than one pollutant.

From a management point of view, airborne pollutants (most of which are listed in Table 1) can rarely be controlled and monitored individually. Doing so would require far too much time and too many resources. One way to narrow down the number of pollutants is to use the 80-20 Rule (Pareto's principle[*]) whereby 80% of pollutants can be controlled in museums by controlling the 20% of pollutants known to be the most significant. Based on this principle, seven pollutants have been designated as key pollutants for museums. The remaining airborne pollutants do not require the same close level of control because their sources, reactivities, and permissible concentrations are equal to or less than those of the key pollutants of the same chemical group. Consideration of a limited number of pollutants also simplifies the risk assessment[*] of the adverse effects of pollutants on the collection. The seven key airborne pollutants — acetic acid, hydrogen sulphide, nitrogen dioxide, ozone, fine particles, sulphur dioxide, and water vapour — are described below.

In this book, the level or concentration of pollutants is expressed in $\mu g\ m^{-3}$. For information on the conversion of concentration units, consult Box 1.

---

1. All terms marked with an asterisk are defined in the Glossary.

BOX 1. CONCENTRATION UNITS

The terms "level" and "concentration" are used as synonyms in this book, although the term "level" can also be used in a more general sense, e.g. the level of protection or the level of risk. In risk assessment, the level is normalized and has the same meaning as concentration.

Two different units can be used to quantify the concentration of most airborne pollutants:
- parts per billion (ppb or ppbv) is a measure of the volume fraction of pollutants in the ambient air; 1 ppb means there is 1 pollutant molecule present in a group of 1 billion air molecules (equivalent to 1×10^{-9})
- micrograms per cubic metre ($\mu g\ m^{-3}$) represents the quantity of a pollutant per unit volume

The two units are related in the following way (21°C and atmospheric pressure of 101.3 kPa):

$$ppb = \mu g\ m^{-3} \times 24.04 \div (\text{molecular weight})$$

For example, to convert 2.5 $\mu g\ m^{-3}$ acetic acid into ppb, multiply 2.5 by 24.04 (the molar volume of a perfect gas) and divide by 60.05 g/mol (the molecular weight of acetic acid). The result is 1 ppb (or 0.0000001%).

As a general rule of thumb, for most pollutants 1 ppb is equivalent to 2 $\mu g\ m^{-3}$. The exact conversions for some key pollutants are provided below.

Key pollutants	Conversion factor
Acetic acid	1 ppb = 2.50 $\mu g\ m^{-3}$
Hydrogen sulphide	1 ppb = 1.42 $\mu g\ m^{-3}$
Nitrogen dioxide	1 ppb = 1.91 $\mu g\ m^{-3}$
Ozone	1 ppb = 2.00 $\mu g\ m^{-3}$
Particles[a] $PM_{2.5}$	
Sulphur dioxide	1 ppb = 2.67 $\mu g\ m^{-3}$
Water vapour[b]	1 ppb = 0.75 $\mu g\ m^{-3}$

a: The concentration of particles cannot easily be converted to ppb due to the different molecular weights of the fine particles.
b: The level of water is usually reported in relative humidity (RH) as a percentage.

A program to convert concentration units can be found on the Internet <iaq.dk/papers/conc_calc.htm> (Ryhl-Svendsen 2001).

Table 1. Sources of airborne pollutants

Airborne pollutants	Indoor and outdoor sources[a]
Amines (RNR)[b]	**Ammonia (NH₃):** alkaline-type silicone sealants, concrete, emulsion adhesives and paints, household cleaning products, visitors, animal excrement, fertilizer and inorganic process industries, underground bacterial activities. **Cyclohexylamine (CHA), diethylamino ethanol (DEAE), and octadecylamine (ODA):** corrosion inhibitor in humidification systems, some vapour corrosion inhibitors. **Aliphatic amines:** epoxy adhesives.
Aldehydes (RCOH) and carboxylic acids (RCOOH)	Aldehydes: **Acetaldehyde (CH₃HCO):** some poly(vinyl acetate) adhesives, wood products. **Formaldehyde (CH₂O):** carpet finishing components, fungicide in emulsion paints, fabric-finishing components, gas ovens and gas burners, natural history wet collections, ozone-generating air purifiers, urea formaldehyde-based adhesive products, tobacco smoke, vehicle exhaust, other combustion. **Carboxylic acids: Acetic acid (CH₃COOH):** acid-type silicone sealants, degradation of organic materials (general) and objects such as cellulose acetate-based objects (vinegar syndrome) and wood products, many emulsion paints, flooring adhesives, human metabolism, linoleum, microbiological contamination of air-conditioning filters, oil-based paints, photographic developing products, some "green" type cleaning solutions. **Formic acid (HCOOH):** degradation of organic materials (general), oil-based paints, wood products. **Fatty acids* (RCOOH):** burning candles, cooking, flooring adhesives, human metabolism, linoleum, lubricant in HVAC systems, microbiological activities from air-conditioning or on objects, objects made of animal parts (including skins, furs, taxidermy specimens, insect collections), oil-based paints, papers, skins, paper and wood products, vehicle exhaust.
Nitrogen oxide compounds (NOₓ)	**Nitric oxide (NO):** agricultural fertilizers, fuel combustion from vehicle exhaust and thermal power plants, gas heaters, lightning, photochemical smog. **Nitrogen dioxide (NO₂):** degradation of cellulose nitrate and same sources as for NO but mainly from oxidation of NO in the atmosphere. **Nitric acid (HNO₃) and nitrous acid (HNO₂):** oxidation of NO₂ in the atmosphere or on a material's surface, possibly the degradation of cellulose nitrate.
Oxidized sulphur gases* (SOₓ or S⁺)	**Sulphur dioxide (SO₂):** degradation of sulphur-containing materials and objects such as proteinaceous fibres, pure pyrite or mineral specimens containing pyrite sulphur dyes, sulphur vulcanized rubbers, petroleum refineries, pulp-and-paper industries, combustion of sulphur-containing fossil fuels. **Sulphuric acid (H₂SO₄):** oxidation of SO₂ in the atmosphere or on a material's surface.
Oxygen (O₂) and ozone (O₃)	**Oxygen:** 21% of the atmosphere. **Ozone:** electronic arcing, electronic air cleaners, electrostatic filtered systems, insect electrocuters, laser printers, photocopy machines, UV light sources, lightning, photochemical smog.
Particles (fine and coarse)	**General:** aerosol humidifier, burning candles, concrete, cooking, laser printers, renovations, spray cans, shedding from clothing, carpets, packing crates, etc. (due to abrasion, vibration, or wear), industrial activities, outdoor building construction, ozone-generating air purifiers, soil. **Ammonium salts:** ammonium sulphate and nitrate: reaction of ammonia with SO₂ or NO₂ in indoor or outdoor environments or on solid surfaces. **Biological and organic compounds:** micro-organisms, degradation of materials and objects, visitor and animal danders, construction activities. **Chlorides:** sea salt aerosol, fossil combustion. **Soot (organic carbon):** burning candles, fires, coal combustion, vehicle exhaust.
Peroxides (ROOR)	**Hydrogen peroxide (HOOH):** degradation of organic materials such as rubber floor tiles, wood products, micro-organism activities, oil-based paints. **Peroxyacetyl nitrate (PAN):** automobile exhaust particularly from alcohol-based fuels, photochemical smog.
Reduced sulphur gases* (S⁻)	**Carbon disulphide (CS₂):** polysulphide-based sealants, fungal growth, rotting organic matter in the oceans, soils, and marshes. **Carbonyl sulphide (OCS):** degradation of wool, coal combustion, coastal ocean, soils, and wetlands, oxidation of carbonyl disulphide. **Hydrogen sulphide (H₂S):** arc-welding activities, mineral specimens containing pyrite, sulphate-reducing bacteria in impregnated objects excavated from waterlogged site, visitors, fuel and coal combustion, marshes, ocean, petroleum and pulp industries (kraft process), vehicle exhaust, volcanoes.
Water vapour (H₂O)	Visitors, water-based paints and adhesives, wet cleaning activities, outdoor environment.

a: Refer to Appendix 1 for quantified data and references.
b: Expressions in parentheses are either the chemical symbol or a common acronym. R represents either a hydrogen bond (H) or a radical such as CH₃(CH₂)ₓ (or even more complex).

TABLE 2. ADVERSE EFFECTS OF AIRBORNE POLLUTANTS

Airborne pollutants	Effects on materials[a]
Amines	**Ammonia:** blemishes on ebonite, corrosion of metal by an ammonium salt, efflorescence on cellulose nitrate. If combined with sulphate or nitrate compounds, it can form a white deposit on the surface of objects. **Other amines:** thought to be responsible for the blemishes on paintings, corrosion of bronze, copper, and silver.
Aldehydes and carboxylic acids	**Acetaldehyde and formaldehyde:** possible oxidation of the aldehyde to carboxylic acids in high RH and/or in presence of strong oxidants. **Acetic and formic acids:** corrosion of copper alloys, cadmium, lead, magnesium, and zinc, efflorescence on calcareous materials such as seashells, corals, limestones, and rich calcium-based fossils, efflorescence on rich soda glass objects, lowering of the degree of polymerization of cellulose. **Fatty acids:** blemishes on paintings, corrosion of bronze, cadmium, and lead, ghost images on glass, yellowing of papers and photographic documents.
Nitrogen oxide compounds	Corrosion of copper-rich silver, deterioration of leather and paper, fading of some artists' colorants.
Oxidized sulphur gases*	Acidification of paper, corrosion of copper, fading of some artists' colorants, weakening of leather.
Oxygen and ozone	**Oxygen with (UV, visible) radiation:** brittleness and cracking of organic objects, fading of colorants. **Ozone:** fading of some artists' colorants, dyes, and pigments, oxidation of organic objects with conjugated double bonds such as rubber, oxidation of volatile compounds into aldehydes and carboxylic acids.
Particles	**General:** abrasion of surfaces (critical for magnetic media), discoloration of objects (especially critical for those with surfaces with interstices that entrap dust, e.g. with pores, cracks, or often micro-irregularities), may initiate or increase corrosion processes due to their hygroscopic nature, or may initiate catalysis forming reactive gases. **Ammonium salts:** corrosion of copper, nickel, silver, and zinc, blemishes on furniture varnished with natural resins and on ebonite. **Chlorine compounds:** increase of rate of metal corrosion. **Soot:** discoloration of porous surfaces (painting, frescoes, statues, books, textiles, etc.), increase of rate of metal corrosion.
Peroxides	Discoloration of photographic prints, fading of some artists' colorants, oxidation of organic objects and pollutant gases.
Reduced sulphur gases*	Corrosion of bronze, copper, and silver, darkening of lead white pigment, discoloration of silver photographic images.
Water vapour	Hydrolysis reaction on organic objects such as cellulose acetate- and nitrate-based objects, some dyes in colour photographs, polyurethane-based magnetic tape or polyurethane foam, photographic gelatine, many type of papers, natural varnishes, flexible (plasticized) PVC. It also increases the rate of other deterioration such as corrosion of metals, efflorescence on calcium-based materials, and photo-oxidation of artists' colorants.

a: Refer to Appendix 2 for quantified data and references.

ACETIC ACID

Acetic acid vapour is a common and reactive indoor-generated pollutant often classified in the chemical group of carbonyls*. It can be released by various products, such as paints, varnishes, poly(vinyl acetate) adhesives, acid-cured silicone, wood products (especially oak and cedar), and some cleaning products. Acetic acid (CH_3COOH) can even be released by some organic objects*. Many lead objects have been damaged when displayed or stored in an enclosure* in the presence of products emitting acetic acid, such as wood products and paints. Visible corrosion on lead can happen in a new display case made of products that emit acetic acid after only 1–3 months of exposure. An example is shown in Figure 1. White powder on the lead coin was caused by acetic acid released by an oak display case. In extreme cases, lead objects can be completely destroyed after a few years, leaving only a white powder. Lead is commonly found as seals on historic documents, small figurines, medals, coins, bullets, modern sculptures, part of composite objects, solder in metal objects, and in metal alloys. In some cases, bronze objects have also deteriorated due to acetic

Figure 1. Corroded lead medals in an oak display case. Courtesy of the Musée du séminaire de Sherbrooke. [A colour version of Figure 1 is available on p. 91.]

acid (Tennent and Baird 1992). Where objects are not in a small enclosure, but are in an open room (except where there are very large surfaces of acid emissive products and inadequate ventilation), there have been very few reported problems caused by acetic acid. The two main strategies to protect acetic acid-sensitive objects are to select non- or low-acid-emissive enclosure products and avoid mixing acid-sensitive objects with acid-emissive objects in the same enclosure.

It is worth mentioning the "vinegar syndrome." This is a typical problem of internal pollutant generation by an object; acetic acid is released by cellulose acetate films (negatives or soundtrack movies) or three-dimensional cellulose acetate objects, especially after reaching a critical phase of hydrolysis* degradation (Reilly 1993).

Formic acid and formaldehyde are two other common indoor pollutants considered harmful by the conservation community. However, they do not have the same wide spectrum of reactivities as acetic acid. The adverse effects of these two compounds can be minimized by avoiding their major sources, such as oil or alkyd paints, and unsealed urea formaldehyde-based glue on wood products in enclosures. Paints formed by oxidative polymerization, such as oil-based, alkyd, and oil-modified urethanes, release both formaldehyde and peroxide vapours during the curing process. A certain amount of the formaldehyde can become oxidized by the peroxide during the first few days after the paint application. Formaldehyde is discussed in the section "Other Pollutants."

HYDROGEN SULPHIDE

Hydrogen sulphide (H_2S), a reduced-sulphur gas with a characteristic "rotten egg" odour, is a key pollutant due to its great capacity to tarnish silver and copper within a year, even in remote areas and at levels far below the human olfactive threshold (1–10 μg m^{-3}). Figure 2 shows the typical tarnishing effect of H_2S. It is also known to darken lead white pigment on paintings. The main anthropogenic* sources of H_2S are the pulp-and-paper and petroleum industries. Outside the urban environment, H_2S is emitted by oceans, volcanic and geothermal activities, marshes, and vegetation (Watts 2000). Inside buildings, elevated levels of H_2S can be found when many people are present. It has been found that a person in a classroom releases about 100 μg of H_2S per hour (Wang 1975).

Figure 2. Tarnished silver-plated copper key ring. Continued cleaning of tarnish compounds will eventually remove the thin silver layer. [A colour version of Figure 2 is available on p. 91.]

Silver objects can tarnish quickly when displayed with waterlogged archaeological objects contaminated by sulphate-reducing bacteria (Little et al. 1998; Green 1992). Carbonyl sulphide (OCS), another common reduced-sulphur compound, is generated mainly in the countryside. It has a lower reactivity on metals than H_2S.

There is a common notion that wool is harmful to silver, and some people have reported fast tarnishing of their silver collection in a room with a wool carpet. On the other hand, silver medals exhibited in display cases with military wool costumes have shown little tarnish after many years. A similar observation was made at the Canadian Conservation Institute (CCI) where silver coupons did not tarnish in a test chamber with a wool sample at about 50% relative humidity* (RH) for more than 8 yrs. In fact, wool does not release significant levels of reduced-sulphur compounds at room temperature. To the contrary, it is rather good at adsorbing sulphur compounds (Crawshaw 1978), which explains the low tarnishing rate of silver in airtight enclosures in the presence of wool. However, the scenario is different with a wool carpet in a room. Wool does tend to release reduced-sulphur compounds when strongly illuminated (Brimblecombe et al. 1992). These compounds, combined with dust and salt with which the carpet may be contaminated and with reduced-sulphur compounds originating from the outside air and visitors, may cause accelerated tarnishing of a silver collection in a carpeted room exposed to direct sunlight.

NITROGEN DIOXIDE

Nitrogen dioxide (NO_2) is the most common compound of the nitrogen oxides group (NO_X) in the atmosphere. NO_2 is responsible for the reddish brown colour above cities, especially during a period of photochemical smog*. It is an important precursor of ozone. NO_2 is rapidly formed in the atmosphere by the action of ozone on nitric oxide (NO). Nitric oxide is the major NO_X emitted by fuel combustion in vehicles (about 50% of emissions), power plants, and industrial activities. Direct emission of NO_2 in the atmosphere accounts for a small fraction of total NO_X emissions. Since the beginning of the industrial age, the emission of NO_X has seriously increased. The national trend of NO_X emission in the United States during the 20th century is shown in Figure 3, which also shows the long-term trends of other pollutants in the United States and a few short-term trends of pollutant levels in Canada and Europe. However, in the last few years in the United States and Europe, there has been a slight decrease in NO_X emissions. Usually, the emission and concentration of outdoor pollutants follow a similar trend for a specific location. This is borne out in some cities in the United States and Canada, where extensive monitoring programs that were begun in the 1980s show a slight decline in the levels of NO_2. Knowing the levels and trends of outdoor pollutants is important because these pollutants infiltrate museums at a fraction of the outdoor levels.

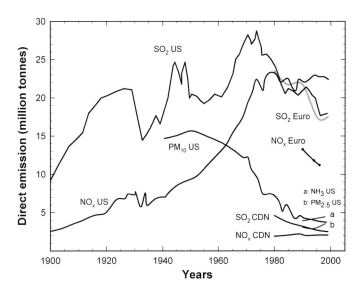

Figure 3. Pollutant emission trends in the United States, Europe, and Canada (EPA 2001c, 2000, 1996; EMEP nd).

Box 2.
Acid deposition

The term "acid rain" was initially used to refer to abnormal acidity found in precipitation. Eventually, to allow for a more global comprehension of the problem of acid compounds falling out of the atmosphere, the term "acid deposition" became more appropriate.

The acid compounds originate mainly from sulphur dioxide (SO_2) and nitric oxide (NO) gases that are emitted as a result of combustion of coal and fossil fuel and smelting activities. In the atmosphere, these gases react with water, oxygen, and other chemicals to form gases and aerosols* (acids, salts, and radicals) of SO_X and NO_X compounds (see Figure 4). They are also frequently adsorbed on other particles. Some reactions will be dominant under solar radiation or in the presence of water droplets (clouds, fog) (Harrison 1996), while others will be dominant at night with the oxidation of NO_2 by ozone.

The sulphate and nitrate aerosol compounds reduce visibility in urban as well as rural areas while they remain in suspension. Eventually they will be deposited on earth in dry or liquid form. About half the compounds in the atmosphere fall back to earth through dry deposition as gases or particles in areas of high pollution close to the sources of emission. It is in this dry form that most airborne pollutants infiltrate the openings or ventilation system of a museum. The other half will fall as wet deposition in the form of acid rain, fog, or snow precipitation. Depending on weather conditions, compounds can be carried hundreds of kilometres before falling through precipitation. During the rain, dry deposited compounds are washed from trees and other surfaces. When that happens, the run-off water adds the acids to the acid rain, making the combination more acidic than the falling rain alone.

Other compounds such as ammonia, carboxylic acids, hydrogen sulphide, hydrogen chloride, and hydrogen fluoride can contribute to the acidity of the atmosphere. These compounds may be from either anthropogenic or natural sources and play a secondary role.

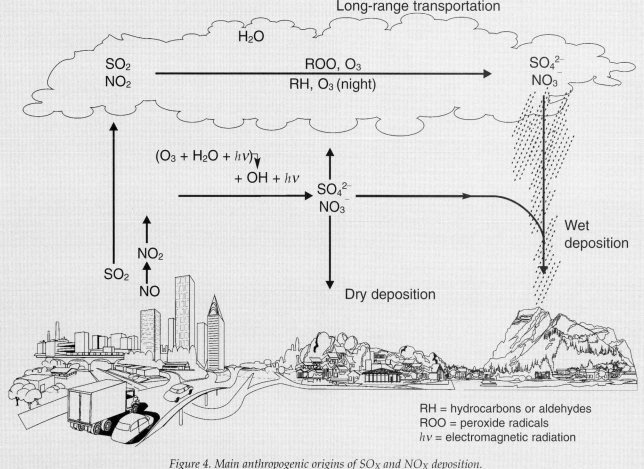

Figure 4. Main anthropogenic origins of SO_X and NO_X deposition.

In the past, NO_X has been known to be the second most important pollutant, after sulphur dioxide (SO_2), responsible for acid rain problems, more accurately known as acid deposition. With the progressive reduction of SO_2 in North America and some European countries, NO_X may eventually become the major precursor of acid rain. See Box 2.

In the atmosphere, a fraction of NO_2 can be further oxidized to its acid form: nitric acid (HNO_3). Both HNO_3 and NO_2 cause artists' colorants to fade and can contribute to the degradation of paper and vegetable-tanned leather. It is also expected that the NO_2 absorbed by objects becomes oxidized to nitric acid, the latter being responsible for most of the ensuing deterioration. Nitrous oxide (N_2O) and NO are not considered directly harmful to collections. A well-known effect of NO_2, as an internal pollutant, is the deterioration of cellulose nitrate-based films (negatives or soundtrack movies) or three-dimensional objects (e.g. combs, barrettes, imitations of turtle shell), the film itself being the source of NO_2 as well as other NO_X gases (Selwitz 1988).

Targeting NO_2 as the key compound of the NO_X family is convenient for monitoring purposes. Many methods detect both NO_2 and NO without discernment. The methods cause the oxidation* of NO to NO_2 on a sorbent*, or they require oxidation of NO by ozone for chemiluminescence analysis.

OZONE

Ozone (O_3) is a strong oxidant* that is normally present in the stratosphere and protects us against intense, harmful ultraviolet radiation. At ground level (within a few hundred metres), it is formed during photochemical smog. Photochemical smog is the result of multiple chemical reactions between nitrogen oxides and hydrocarbons, and their oxygenated derivatives in the presence of sunlight, as shown in Figure 5. The level of ozone increases after the morning traffic rush which is a source of ozone precursor compounds. With stronger sunlight, the ozone level reaches a peak in the afternoon. During the photochemical process, other harmful pollutants such as acids and fine particles are formed. In parallel with the presence of precursors, meteorological factors (i.e. solar radiation, rains, wind speed, and wind direction) strongly influence the formation of ozone at ground level. High surface temperatures from strong solar radiation cause greater effects on the peak level of ozone

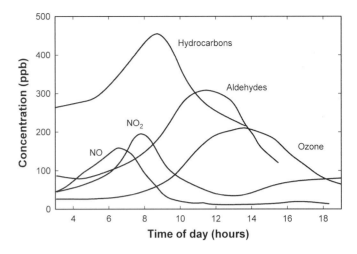

Figure 5. Daily evolution of smog (adapted from Manahan 1994).

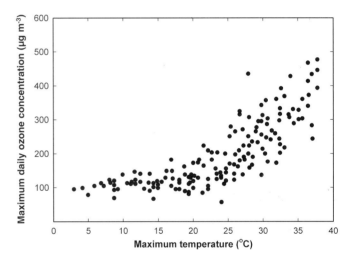

Figure 6. Ozone levels versus temperature in New York City. The values on this graph represent the maximum daily ozone levels and the maximum daily temperature in New York City from May 1988 to October 1990 (NAST 2001).

than changes in the precursor concentration. Figure 6 illustrates this correlation between ozone levels and temperature for New York City. In Canada, the United States, and central Europe, many smog periods happen during the summer, while in coastal European countries such as England and The Netherlands the peak ozone levels happen in spring (EC 1999; Hjellbrekke 2000). In Canada, the Windsor–Quebec City corridor has the highest levels of ozone. This is due to the high population density and the industrialization of the corridor, and to the dominant southwest winds that carry ozone precursors, especially from the immediate area south of the Great Lakes. The transport of ozone precursors in the atmosphere causes high levels of ozone in remote areas where there are no major human activities. Over the last decade, ozone level trends at ground level have been fairly stable in the United States and Europe but have seemed to increase slightly in Canada, despite the efforts of both the American and Canadian governments to reduce the emission of two major ozone precursors: hydrocarbons and NO_2 (EPA 2001; EC 1999; Hjellbrekke 2000).

Inside buildings, the main sources of ozone are electrostatic precipitators* in the heating, ventilating, and air-conditioning (HVAC) system*, electronic air cleaners (ozone generators), and photocopiers. In theory, ozone can attack materials by breaking apart any double bonds between carbon atoms. The degradation of vulcanized natural rubbers under stress and the fading of artists' colorants are the most studied phenomena. Even though organic objects have a high potential for deterioration, little quantitative data exist to support the widespread assumption that they are significantly altered by ozone under normal conditions.

FINE PARTICLES

It is common to characterize particulate matter (dust) in terms of diameter. This property is important because it determines behaviour and control. Particles have been divided into a few groups based on their aerodynamic diameter*. For the control of pollutants, the fine particle ($PM_{2.5}$: suspended particle matter having an aerodynamic diameter equal or less than 2.5 μm) and the coarse particle (PM_{10}: aerodynamic diameter between 2.5 and 10 μm) are commonly used as indicators. Figure 7 shows size and mass distribution, formation processes, and the deposition velocity* of different particles. Sulphate and nitrate compounds, organic carbon, crustal* materials, and salts are the major harmful compounds from fine particulate matter ($PM_{2.5}$) from

outdoors (EPA 2001). They are discharged directly into the atmosphere or formed in the atmosphere from secondary reactions. Canada and the United States show decreasing levels over the last 15–20 yrs (EPA 2000; EC 1999). Because smaller particles can lodge in the smaller interstices on an object's surface, $PM_{2.5}$ is the most harmful particle size, and its control will also reduce significantly the levels of gaseous pollutants which tend to be grouped by nucleation or be adsorbed by the particle. As shown in Figure 7, fine particles having a diameter between 0.05 and 2 μm tend to accumulate in the environment due to their low deposition velocity. They can be in the ambient air* for a few days. Due to their small size, they are also the most challenging particle size to control. Any attempt to control the level of $PM_{2.5}$ must consider *a priori* the control of levels of PM_{10} and some super coarse particles (>10 μm) which still contain some potentially reactive compounds, such as combustion residues, human danders, and microbiological specimens. Compared to

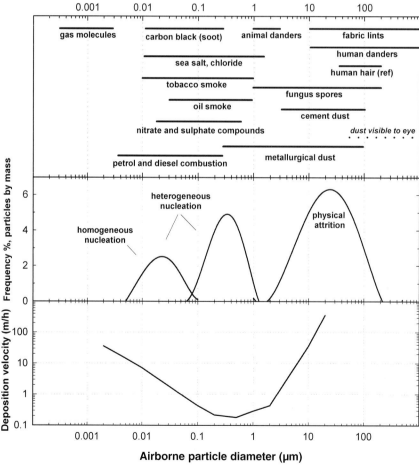

Figure 7. Dust distribution. Particle diameters are taken from various sources. The diameter of a human hair is shown for reference only. Deposition velocity adapted from Slinn et al. (1978).

fine particles, particles larger than 10 μm have a short suspension time*. They are found close to their sources if not carried by strong winds. Fine particles are particularly damaging, because they discolour or soil surfaces. Soiling changes the visual perception of objects. The more fragile, porous, or altered the surfaces, the more difficult they are to clean. Any control strategy designed to maintain low levels of particles is beneficial to objects since cleaning fragile or porous objects can be difficult. Object cleaning is a delicate process that requires time and trained conservators. The ivory sculpture and delicate First Nations hat made of feathers and down shown in Figures 8 and 9 are two good examples of objects that are challenging to clean. Another example would be filamentous mineral specimens, which are probably impossible to clean using conventional methods.

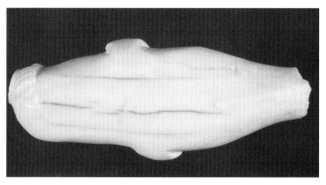

Figure 8. Inuit ivory sculpture with soot encrusted in the cracks. [A colour version of Figure 8 is available on p. 92.]

Figure 9. Native American feather headdress. The fragility of the materials makes this object very hard to clean.

Deposition of hygroscopic, oily, or metallic particles on a surface can initiate or accelerate deterioration as well as the formation of harmful compounds, such as acids. Except for particles generated by cooking activities in a museum's cafeteria or burning combustibles (candles), most indoor-generated particles are composed of soil, dust, and carpet and cloth fibres. Fibres are not generally considered to have direct adverse effects on a collection, with the exception of magnetic media such as audio and video tapes where abrasive dusts are an issue during handling and playing. Dust accumulation can also provide an attractive foraging place for insects and mould. Another adverse consequence from a wider viewpoint is the impact of the perception by visitors, including potential donors, that there is a basic lack of care for the collection.

Filtration of outdoor fine particles should be considered as an important control strategy. However, not all particle sizes are evenly controlled. The most dense and the biggest indoor-generated particles (probably bigger than an aerodynamic diameter of 50 μm) do not easily reach filters of a HVAC system. Their suspension times (a few seconds) are too short to be trapped by the air filter system and they fall with gravity. These dust particles can eventually resuspend with air movement from human activity. Dust can also be released by objects and products due to vibrations and dimensional changes from RH fluctuations. Periodic vacuum cleaning is needed and the vacuum cleaner should have a high-efficiency filter. Airtight enclosures and those with a positive pressure system are two good options to prevent dust deposition on objects.

SULPHUR DIOXIDE

Since 1900, most energy consumed by industries, transportation, and heating in North America has originated from fuel or coal combustion. Figure 3 shows the rise and fall of sulphur dioxide (SO_2) levels in the United States. SO_2 is the main compound responsible for acid deposition (see Box 2). In areas with high levels of SO_2, acid precipitation has had serious negative impacts on building structures, outdoor monuments, and the overall ecosystem. Many leather books stored in urban archives from the start of the industrial age have been severely damaged (Figure 10). It is also expected that SO_2 oxidizes to its acid form (H_2SO_4) on objects in the presence of metallic ions or salts. Fortunately, the regulation of SO_2 emissions

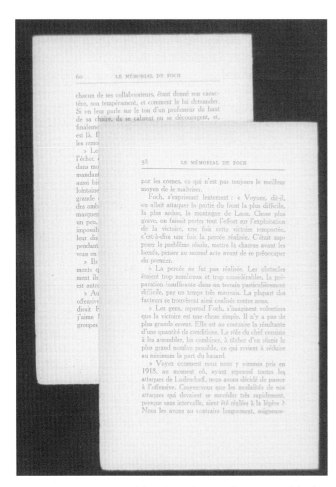

Figure 10. Browning of the edges of a page from a French book printed in 1929. The pollutants (mainly SO_2) were adsorbed on the edges and are slowly diffusing to the centre of the page. The pH of the darker zones is about 3.5 and the pH in the centre of the pages is 6.2. [A colour version of Figure 10 is available on p. 92.]

in the 1970s greatly reduced its atmospheric level. Today, power plants based on coal, and oil combustion in the United States and in Europe are the major sources of SO_2, followed by industrial processes and transportation (EPA 2000; Clean Air 2000; EEA 2001). Only small quantities of SO_2 come from gasoline-fuelled motor vehicle exhaust. In Canada, industrial activity, specifically metal smelting, is the main source of SO_2 (Stadler-Salt and Bertram 2000; MOE 1999). According to the World Resource Institute, emissions of SO_2 should remain stable during the next few decades in Canada, the United States, and Europe if no major environmental policies change. Unfortunately, an important increase in SO_2 is expected in Asian countries where economic development is increasing without being followed by strict pollutant emission regulations (WRI nd).

Material inside enclosures (i.e. proteinaceous materials, sulphur-vulcanized rubbers, and oxidizing sulphides in geological specimens and some dyes) are sources of sulphur compounds. While damage to objects has been attributed to them, the sulphur compound gases generated by these objects and products in enclosures have not been monitored closely.

WATER VAPOUR

Water (H_2O) is included as a key airborne pollutant even though there are well-established guidelines for RH levels for museums to prevent physical deterioration caused by incorrect levels (too dry or too humid) or by excessive fluctuations. The action of water relates to both physical and chemical damage. Water vapour can directly damage, by hydrolysis, cellulose-based materials* which are usually an important part of collections. Materials that are sensitive to the hydrolysis action of water vapour include cellulose acetate and nitrate (Figures 11,

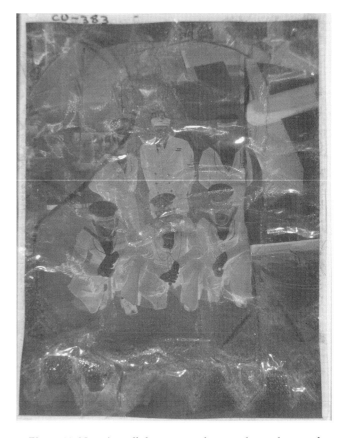

Figure 11. Negative cellulose acetate sheet at advanced stage of degradation. The anti-curl layer and the emulsion layer have turned yellowish, and are 10% channelled and 100% blistered. The film base is yellowed and very brittle. It has completely degraded and converted from the ester to cellulose (i.e. regenerated cellulose). [A colour version of Figure 11 is available on p. 93.]

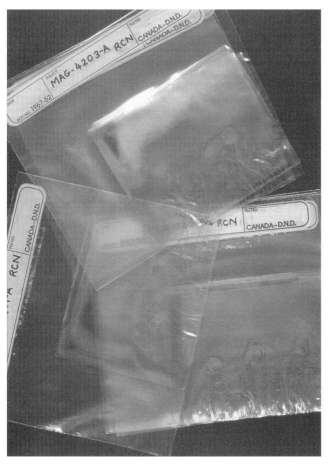

Figure 12. Negative cellulose nitrate sheets at advanced stage of degradation. The sheets have turned brown-yellow. [A colour version of Figure 12 is available on p. 93.]

Figure 13. Highly brittle cellulose nitrate comb made in the 1960s.

12, and 13), especially in the form of thin sheets or rolled films, papers, and polyurethane-based magnetic tapes. Water vapour also greatly influences the deterioration processes caused by other pollutants (see section "Factors Affecting the NOAEL and LOAED"). Based on the 80-20 rule and the great impact of water vapour on the collection, it is inevitable that it would be a key pollutant.

An increasing number of visitors in a poorly ventilated room can increase water vapour levels inside buildings. Inside an enclosure, newly applied liquid products* (such as water-based paints or adhesives) can elevate humidity levels. The control of water vapour in the room or in the enclosure is not always obvious. However, the drier the better (as low as 20% RH) in terms of preserving many objects, such as metals, shells, paper-based materials, and many plastics. Often, target levels must consider the RH specification of a composite collection or a composite object, not forgetting the historical average RH levels in the building. For example, antique furniture displayed for a very long time in a humid historical house cannot easily be moved to a drier environment without risk. Important dimensional changes of the furniture components will occur during the acclimatization to the new environment.

Another problem related to high RH is the possibility of mould. Figure 14 shows the number of days it takes, at room temperature, for the onset of mould at various RH levels. The lowest humidity shown to produce mould very slowly is 60%. At 70% RH, it takes months for mould to grow.

The issue of incorrect RH is touched on only slightly in this book as it is extensively covered elsewhere (Michalski 2000; ASHRAE 2003; Erhardt and Mecklemburg 1994).

OTHER AIRBORNE POLLUTANTS

Besides the seven key airborne pollutants already described, other pollutants may need

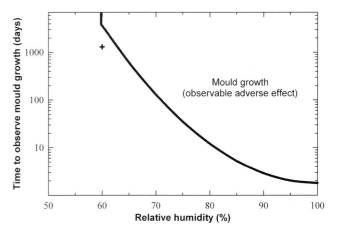

Figure 14. Time required for visible mould growth, assuming a highly susceptible material at about 25°C and RH that has climbed (not fallen) to these values. The cross indicates no growth in 1300 days at 60% RH (Michalski 2000).

to be investigated and controlled, if they are present in unusually high levels or if the museum houses a collection that is particularly sensitive to them. Tables 1 and 2 can be consulted for a preliminary risk assessment of these specific cases. Keep in mind that adding new pollutants to the environmental control specifications will increase costs and resources associated with long-term monitoring programs. It may not be worthwhile to monitor a pollutant extensively when a designated "key" airborne pollutant of the same chemical group is already controlled: the control of one or two key airborne pollutants will often control the pollutant of concern. However, the identification of potential pollutants in new products or in objects remains an important task. By identifiying incompatible products, control strategies can be adapted to prevent or minimize deterioration of the collection. Also, the effects of some gases or vapours on collections are still not clear or are the object of debate among researchers in the field. The rationale for the exclusion of ammonia, carbon dioxide, formaldehyde, oxygen, and volatile organic compounds as key pollutants is explained below.

AMMONIA

The main outdoor sources of ammonia (NH_3) are livestock agriculture and fertilizers (EPA 2000). In an indoor environment, a high level of ammonia can result from a large number of visitors (Dahlin et al. 1997). Periodic peak levels of ammonia may occur if cleaning products used for housekeeping contain this pollutant (e.g. window-cleaning fluids). In indoor environments, most damage from ammonia is associated with ammonium salts which cause corrosion or a whitish deposit on an object's surface. The salts are formed by nucleation of ammonia with other compounds, such as sulphate or nitrate compounds. By controlling the levels of SO_2, NO_2, fine particles, and RH, the adverse effects of ammonia should be negligible. However, in some cases ammonia reacts directly with a component of an object and forms a salt — such as some cellulose nitrate objects as shown in Figure 15.

CARBON DIOXIDE

Carbon dioxide (CO_2) is another gas that is monitored more for its human health impact than for the preservation of collections. Monitoring CO_2 levels in a building operating an HVAC system helps establish the optimal human comfort level with a minimum rate of air ventilation. It has been shown that CO_2 in its gas phase has very little adverse effect on an indoor collection even at high levels

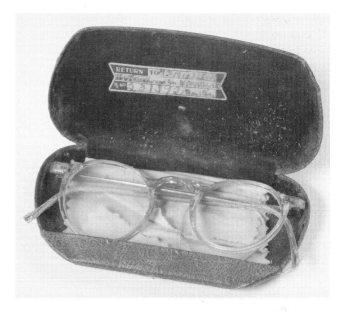

Figure 15. Eyeglasses with a cellulose nitrate frame. (Top) The glasses in their leather case; (bottom) a close-up of the cellulose nitrate frame showing the presence of ammonium nitrate filaments.

as long as RH is below 75%. Much atmospheric research underlines that an increasing amount of CO_2 is an important consequence of climate change. More information can be found on the following Web sites: *Climate Trends and Variations Bulletin* from Environment Canada (EC, nd), *Climate Change* from the NewScientist (nd), and *Global Warming* from the Cooler Heads Coalition (nd).

FORMALDEHYDE

This aldehyde has been, perhaps unfairly, extensively publicized for its harm in conservation. Since the 1970s, with the increased use of wood product panels and urea formaldehyde insulation foam, formaldehyde (CH_2O) has been known for its adverse effects on human health. In a museum context, some people have observed the corrosion of lead objects in enclosures (Leveque 1986) with urea formaldehyde-based wood products. Curiously, it is hard to reproduce this problem in laboratories without extreme conditions (Tétreault et al. 2003; Thickett 1997). With the exception of lead, very few case studies support the harmfulness of formaldehyde to collections in a museum environment. By comparison, acetic acid seems to be a more aggressive indoor organic compound for a larger range of objects, but does not show serious harmful effects on human health. Formaldehyde often has a minor role in the corrosion of metal (Tétreault et al. 2003; Thickett et al. 1998), and the same minor role is also expected in the efflorescence of calcium carbonate-based materials, such as eggs or shells. The oxidation of formaldehyde into formic acid (HCOOH) is negligible in the atmosphere without the abnormal presence of oxidants. However, it is possible that some oxidation of formaldehyde to formic acid happens on a material's surface; this can be stimulated by deposits or the morphology of the object's surface, and can lead to the growth of formate or carbonate compounds. This possibility has not been demonstrated clearly. With North American regulations on the amount of formaldehyde in products, control strategies for ambient air levels of acetic acid will generally control the level of formaldehyde.

OXYGEN

Oxygen (O_2) is naturally part of the atmosphere. Without necessarily being the initiator, oxygen is involved in many deterioration processes of organic materials, such as some colorants, polymers, cellulose-based objects, and skins. The deterioration is caused by the oxidation of a compound after being photo-excited by UV or visible radiation. The resulting oxidation leads to physical changes, such as brittleness and cracking, as well as chemical changes, such as yellowing and fading. In the presence of moisture, metals such as iron will rust. Obviously, there are many advantages in having an oxygen-free environment. Up to now, low oxygen or anaerobic environments have been used mainly for disinfestation and for long-term storage of individual objects in airtight bags. There are few case studies of the use of low-oxygen environments for large enclosures. This is partly due to problems related to airtightness, maintenance costs, and access. Because of the limited use of low-oxygen environments and the unnecessary need to include oxygen in basic monitoring campaigns, oxygen has not been classified as a key pollutant.

VOLATILE ORGANIC COMPOUNDS

Volatile organic compounds (VOCs) are a class of chemical compound mixtures that contain one or more carbon atoms and tend to evaporate at room temperature. They are solvents released by such things as cleaning products, paints, and wood products. People tend to assume that pollutants that are harmful to human health are also harmful to objects, but high levels of VOCs do not necessarily mean a high risk to a collection. Most individual VOCs are not harmful to objects, with the exception of aldehydes and carboxylic acids as already identified in Table 2. However, even a low level of VOCs in an enclosure can be dangerous if a high proportion of harmful pollutants is present. Minimizing the release of VOCs from different consumer products and from industrial and combustion processes is important because some of them, such as hydrocarbons, contribute to the formation of ozone outdoors at ground level. Efforts have been made to reduce VOC emissions in the United States since the 1970s, and results have been encouraging (see Figure 3).

QUANTIFICATION OF THE EXPOSURE–EFFECT RELATIONSHIP

2

To establish defensible exposure guidelines, it is essential to determine the quantitative relationship between a given pollutant and its effect on materials. Two approaches have been chosen to determine the exposure–effect relationship: the no and lowest observed adverse effect levels (NOAEL and LOAEL) of an airborne pollutant surrounding an object, and the doses* (concentration x length of time). These approaches are the foundation of contemporary quantitative risk assessment (ACS 1998). The notion of deterioration of objects needs to be defined precisely in the context of these concentrations and doses. First, deterioration is a complex function involving many environmental parameters. Fortunately, in indoor environments, many parameters can be considered as pseudo-constant. This assumption simplifies the determination of the effect over time of a single parameter, such as the main airborne pollutant. The second point relates to which chemical and physical characteristics of the objects are the most significant expression of an adverse effect or loss of value. This is a subjective notion and depends on the pollutant–object system*. The characteristics and the method used must be the most reliable and meaningful for the decision-makers in the museum. For a comprehensive evaluation of a specific pollutant–object system, more than one characteristic may need to be monitored. More information on these two issues is reported in Boxes 3 and 4. This chapter explains the approaches based on the concentration or dose of pollutants where no or minimal adverse effects are observed, as well as the factors affecting their values. In the following chapter, these approaches will serve as assessment tools to determine the risk to a collection of the presence of pollutants.

DETERMINATION OF NOAEL AND LOAED

A no observed adverse effect level can be defined as the highest level of a pollutant that does not produce an observable adverse effect on a specific chemical or physical characteristic of a material in a specific experimental set-up (analytical method, exposure time, temperature, RH, etc.). To determine the most reliable NOAEL, the materials should be exposed to different concentrations of pollutants. The NOAEL is the level just below the concentration where an adverse effect or a predefined loss is observed. Figures 16 and 17 provide two examples of the

---

BOX 3.
DETERIORATION IS A COMPLEX FUNCTION

Specific airborne pollutants can react with specific materials in an object to cause deterioration at a given rate. This rate can be expressed as the summation of many chemical and physical parameters.

The rate of deterioration depends on the nature of the object (including the state of the materials and surface morphology), level of the specific airborne pollutant, level or amount of other airborne pollutants such as oxidants, sorbed water, and deposited particles (salts, oily and metallic compounds), temperature, radiation, and air movement.

For risk assessment of the adverse effect of airborne pollutants on materials to be feasible, some assumptions must be made to simplify the complex function of deterioration. The prediction of adverse effects can be helped by reducing the number of variables, which can be accomplished by treating some of them as "pseudo" constant. For example, visible and ultraviolet radiation can be kept low and the temperature kept between 20 and 30°C. Fortunately, museums already avoid extreme environmental conditions, such as direct exposure to sunlight, large fluctuations of RH, and high temperatures. Ideally, the complex function of deterioration can be simplified to the point where the only variable parameters are the level of the main pollutant for a specific material and the exposure.

The effect of environmental parameters on collections reinforces the need for global, balanced preservation strategies. It would be wrong to pretend that acute control of two or three environmental parameters constitutes a long-term preservation plan. Also, it has to be recognized that although fire, water (e.g. damage from broken pipes), theft, and mishandling are intermittent, they can cause a great loss of value in a very short time compared to the loss due to airborne pollutants.

BOX 4.
LOSS OF VALUE

The notion of "loss of value" or "damage" is subjective (i.e. can vary depending on person and time). For example, an artwork created on paper that has yellowed may look normal or acceptable to some people since no information seems altered, but others will be bothered by the yellowing and view it as a sign of net loss of initial properties. For scientists, even before the paper turns yellow, measurable decreases in the chemical and physical properties can provide information on the long-term behaviour of the paper.

Some terms merit clarification. The deterioration or degradation of an object can be defined as a change in its material state. On the other hand, a damage or loss of value is the consequent loss of attribution or value (aesthetic, scientific, historic, symbolic, monetary, spiritual, etc.) of the object. In risk analysis in the fields of health, safety, and environmental policy, the term adverse effect is commonly used. This term refers to an abnormal or undesirable effect on objects, and usually reflects a change in a specific chemical or physical characteristic of the material.

The quantification of adverse effects is not easy due to the lack of standard protocols, incomplete documentation, or incomplete or indecisive understanding of the exposure–effect relationship. Still, some kind of quantification must be available for the decision-makers. A change in colour of a paper document is, in some cases, a simpler issue than all the different chemical and physical properties (e.g. pH, degree of polymerization, folding endurance, and tensile strength) used to characterize damaged papers. The concepts of NOAEL and LOAED are well adapted for quantifying changes in chemical or physical properties of objects, but not necessarily adequate for the overall evaluation of loss of value. Dyes from a textile may be completely faded due to past exposure to pollutants and light, and from the point of view of the colour itself this textile may look

like a total loss. But the object still remains. At this point the control strategy should shift to focus on maintaining the physical strength and cleanliness of the textile.

The impact of an adverse effect on the value of an object will vary depending on the state of the damaged material and its role in the object. In some cases, the loss of value can depend on how easy it is to recover the object. For example, it is easier to retrieve the aesthetic appearance of a tarnished silver drinking glass without much design than a tarnished tiny silver ornament on an antique costume.

Some types of deterioration are changes to the original state of the material that are not considered to be adverse effects by some people. In fact, in some cases they are considered to improve the artwork. The best example of this is the natural formation of patina on bronze sculptures. This is not usually made by the artist, but probably anticipated. For the public, patina gives a sense of historic value. For conservation scientists, it is an alteration or a corrosion that usually provides protection against further corrosion if the corrosion layer is made of low-soluble compounds. A second example is the small lead bullets of a whitish or greyish colour that are often found on display in military museums. Some of these bullets may have corroded while they were still in the field, but most of them probably corroded inside the display case or the wooden storage cabinet. Some curators are so used to seeing them this way that it looks normal to them.

As yet there is no quantitative way to assess loss of value, so this book focusses mainly on the loss of physical or chemical properties of the materials. Work is still needed to define the most critical adverse effects for various materials.

For more information on the issues of value and damage, see Ashley-Smith (1999: pp. 81–119) and Michalski (1994c).

---

determination of NOAEL of acetic acid for metals (lead and zinc, respectively). Metal plate samples were exposed to different levels of acetic acid vapour, and the weight gain was monitored monthly. After a few months of exposure to acid, a NOAEL could be determined for each metal. As expected, lead shows a higher vulnerability to acetic acid than does zinc, with respective NOAELs of 400 and 22 000 μg m^{-3}. The NOAEL approach relies on thermodynamic

limitations (Brimblecombe 1994). However, a NOAEL found experimentally can be quite different from theoretical calculations. So far, few NOAELs have been reported for pollutant–object systems; this may be due to the difficulty in monitoring minor changes over long periods, or because the NOAEL approach is not applicable to some pollutant–material systems*. When the NOAEL cannot be determined with confidence, the lowest observed adverse effect

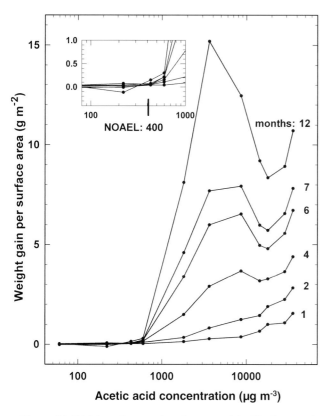

Figure 16. Weight gain over time for untarnished lead exposed to various concentrations of acetic acid at 54% RH. The insert is a close-up of the concentration of acetic acid at which an adverse effect is observed (Tétreault et al. 1998).

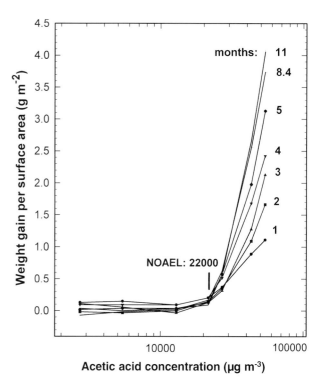

Figure 17. Weight gain over time for untarnished zinc exposed to various concentrations of acetic acid at 54% RH (Tétreault 1992a). Corrosion takes place under the same conditions as in Figure 16.

dose (LOAED) can be used to quantify the exposure–effect relationship.

LOAED is the cumulative dose (LOAEL x time) at which the first signs of adverse effects are observed. With only one concentration, the LOAEL is this value by default. Figure 18 illustrates how to determine the LOAED using the example of the fading of basic fuchsin (a green pigment) as a result of exposure to 450 μg m^{-3} (LOAEL) of sulphur dioxide. The critical exposure time is the time required to observe the adverse effect. This period is found at the intersection of the fading curve and the significant loss: $\Delta E = 2$. Colour changes of ΔE greater than 2 are commonly perceived as adverse effects for colorants. The LOAED is 450 μg m^{-3} x 11 days. A dose of 14 μg m^{-3} yr is obtained after normalization. For other types of experiments, a change of 5% in the initial chemical and physical properties has often been used. With an experiment using different concentrations of pollutants, the lowest concentration causing detectable deterioration should be chosen to determine the dose. The LOAEL used to determine

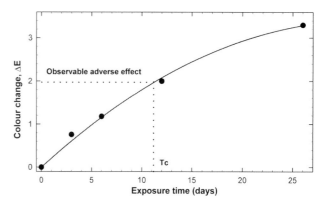

Figure 18. Determination of the LOAED of sulphur dioxide for basic fuchsin. The adverse effect is determined by a change in colour using the CIE L* a* b* colour system. Pigment was exposed to 450 μg m^{-3} of SO$_2$ at 50% RH and 21°C. Tc = critical time to reach an observable adverse effect. LOAED = SO$_2$ concentration x Tc = 450 μg m^{-3} x 11 days x (1 yr/365 days) = 14 μg m^{-3} yr.

the LOAED can be a reliable estimate, since it is usually calculated on the linear portion of the deterioration pattern. Figure 19 shows the fading of basic fuchsin by SO$_2$ over a short range of concentration (450–1300 μg m^{-3}) and exposure

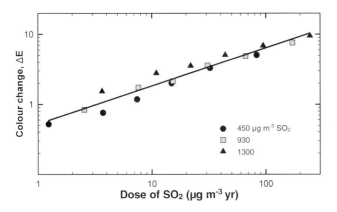

Figure 19. Discoloration of basic fuchsin by various doses of SO₂ at 50% RH, 21–23°C, in the dark. Doses were obtained from three different concentrations and exposure periods varying from 1 to 67 days. Colour measurement was done with a Minolta CR200 chromometer using the CIE L* a* b* colour system. ΔE compares the colour of the sample with its initial colour (Tétreault and Lai 2001).

period (1–67 days). The change in colour as a function of the dose follows the linear reciprocity principle*. The linear reciprocity principle is also

well characterized for the corrosion of silver and copper under hydrogen sulphide and carbonyl sulphide environments over 2 or 3 orders of magnitude* (Graedel et al. 1985). This demonstrates that, to some extent, the linear reciprocity principle can be used for the approach of LOAED.

The linear reciprocity principle allows for the estimation of the time required to observe an adverse effect on a material at lower pollutant concentrations. For example, copper has a significant tarnish layer after being exposed to an average level of 1 μg m^{-3} of hydrogen sulphide for 1 yr. Based on the reciprocity principle, it takes about 10 yrs to obtain the same deterioration at a level of 0.1 μg m^{-3}. Figure 20 illustrates a good correlation between fine particle soiling and the time required to observe visible signs of soiling over 3 orders of magnitude of fine particle concentrations. The approach of LOAED has some potential for the assessment of the deposition of particles and for the adverse effects of a gaseous pollutant on materials.

While some pollutant–material systems follow a linear reciprocity principle according to an experiment, this reciprocity is not usually linear over an extended range of doses. The deterioration versus the dose can follow auto-retardant patterns where fast deterioration is observed at the beginning and is reduced progressively over time. This type of reciprocity pattern has also been observed for light fading of some colorants (Saunders and Kirby 1996). The concept of LOAED can also be applied to other agents of deterioration. Appendix 8 provides a comprehensive application of LOAED associated with light fading.

It should be emphasized that LOAED allows only estimations. Caution is required in using the reciprocity principle over a few orders of magnitude

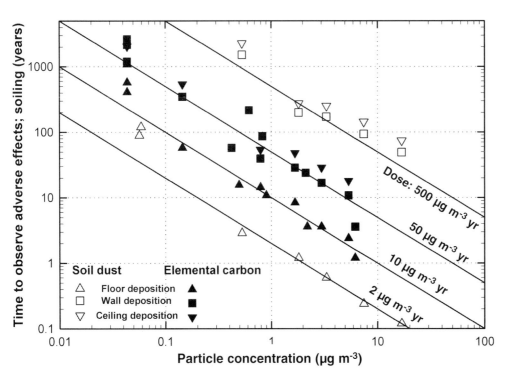

Figure 20. Estimated time (years) before soiling due to soil dust and elemental carbon deposition is observed. Adapted from a model. Because elemental carbon is the most important fine particle, PM₂.₅ will give the same darkening (less than 5–20%) as elemental carbon alone. The model assumes homogeneous air distribution (Nazaroff et al. 1993; Bellan et al. 2000).

of an exposure period due to increasing uncertainties. Figure 21 shows other examples of reciprocity of some materials at different levels of water vapour. Accelerated ageing experiments* show an acceptable linearity of the dose for the least stable dyes on photographs and for cellulose acetate film in the range of 20–80% RH. However, there is a weak correlation for the polyurethane videotape based on the extractable compound method. Physical tests, such as the peel force, seem more adequate for lifetime estimations of polyurethane tapes.

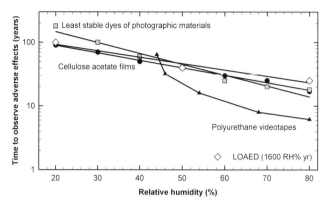

Figure 21. Estimated time (years) before adverse effects are observed on objects at 21°C and various RH levels. The objects and critical adverse effects chosen were: cellulose acetate films, 0.5 free acidity; early 1990s colour slides, negatives, and prints, 30% (dark) fading of least stable dyes; and high-grade VHS urethane tape, 12% hydrolysis. The line joined by the white diamonds shows the linearity of the LOAED of water vapour applied to cellulose acetate (80% RH x 20 yrs = 20% x 80 yrs). For the conditions below the various curves, no adverse effects are observed. Data were obtained by accelerated ageing (Reilly 1993, 1998; Van Bogart 1995).

For a given pollutant–material system, it is possible to predict the rate of observation of the first signs of an adverse effect on the material (i.e. an adverse effect every 1 or 10 yrs) by knowing the average level of a pollutant. If the average concentration of the pollutant changes significantly, the interval of the observed adverse effect will change proportionally. There is no justification for very low concentrations of pollutants where processes such as oxidation, hydrolysis, and thermal degradation become the dominant factors of deterioration. For example, it is unnecessary to specify very low levels of sulphur dioxide (<0.1 μg m^{-3}) for the long-term preservation of acidic papers, because water vapour and temperature are the dominant agents of

deterioration*. What the acceptable rate of deterioration for a material really is in the context of the preservation of the object is a complex decision, and will be discussed in Chapter 5.

A critical review of detailed in situ observations, combined with laboratory studies, has provided substantial quantitative information on the adverse effects of pollutants on materials. An extensive list of NOAELs and LOAEDs extracted from the literature is given in Appendix 2 and summarized in Table 3. The bottom of the table is a reciprocity scale that extrapolates LOAEL over time for a specific LOAED. Some materials are more vulnerable than others. However, the possibility of a rapid, observed adverse effect will depend on the level of the pollutant. More information on the notion of vulnerability is covered in Box 5.

BOX 5.
VULNERABILITY

An object can be made of more than one material, and the vulnerability or sensitivity of the object will depend on the vulnerability of each material. The risk of adverse effect is a different issue.

A material is considered vulnerable or sensitive to a specific pollutant if it reacts to a low NOAEL or LOAED of that pollutant. But the object can be vulnerable or sensitive without necessarily being at risk of deterioration, because risk depends on the level of pollutants or the probability of an adverse consequence.

For example, the NOAEL of acetic acid is high for lead (400 μg m^{-3}) (i.e. lead is not vulnerable or sensitive to acetic acid). In an open room where the levels of acetic acid are usually well below its NOAEL, lead is not at risk. However, in an enclosure having acid-emissive products, it is at risk. In contrast, the LOAED of hydrogen sulphide is low for silver (0.1 μg m^{-3} yr) (i.e. silver is highly vulnerable). In an open room silver objects are likely to tarnish easily in 6–12 months, whereas in an enclosure the adverse effect will take much more time to be observed. The difference between lead with its high NOAEL compared to silver with its low LOAED is that lead may never corrode if the acetic acid is kept below its NOAEL, while shiny silver will eventually tarnish significantly: it is just a question of time.

TABLE 3. QUANTIFIED ADVERSE EFFECTS OF AIRBORNE POLLUTANTS ON VARIOUS MATERIALS[a]

Dose (µg m⁻³ yr)	0.1	1	10	100	1000	10 000
Acetic acid				**lead 400 µg m⁻³** [b]	**tarnished lead 3000 µg m⁻³** copper 1000 **cotton rag paper 4000 µg m⁻³**	**zinc 20 000 µg m⁻³** **calcium-based mat 10 000 µg m⁻³**
Carbonyl sulphide			copper 30	silver 500		
Formaldehyde				copper 600	zinc 6000	
Formic acid				**lead 200 µg m⁻³**	**copper 8000 µg m⁻³**	
Hydrogen peroxide		black-and-white photo 1				
Hydrogen sulphide	**0.1 silver**	copper 1				
Nitric acid	**0.1 sensitive colorants**					
Nitrogen dioxide		sensitive colorants 1	micrograph film dyes 100			
			50% faded sensitive colorants 10 cotton rag paper 10 newspapers 50 copper 50		silver image 2000	
Ozone		sensitive colorants 1 alizarin crimson (AC) 2 AC in acrylic binder 8 wet cotton 3	AC covered by acrylic film 50 AC in acrylic binder covered by acrylic film 60 dry cotton 80	micrograph film dyes 300		
	<-0.005 stressed vulcanized natural rubber					
Particles (2.5 µm)			horizontal surfaces 10 vertical surfaces 50			
Sulphur dioxide			bleached kraft papers 10 40 vegetable-tanned leather sensitive colorants 10 newspapers 50 cotton rag paper 10 copper 50			
Water vapour				**mould growth 60% RH**	cellulose acetate 2000% RH yr most sensitive photographic dyes 2000% RH yr	

Max. allowable levels of airborne pollutants (µg m⁻³) to maintain a low risk of damage to materials for the indicated exposure periods (based on LOAED):

1 yr exposure:	0.1	1	10	100	1000	10 000
10 yrs:	0.01	0.1	1	10	100	1000
100 yrs:	0.001	0.01	0.1	1	10	100

a: NOAEL and LOAED are estimated based on the most reliable data from Appendix 2 at 50–60% RH and 20–30°C, with a clean collection. If these conditions are not met, the values will change. For example, if the average RH is much above 60%, the preservation target (in yrs) could be reduced by an order of magnitude.

b: Values in bold are the NOAEL of the pollutant and are expressed in µg m⁻³; all other values are the LOAED and are expressed in µg m⁻³ yr.

Factors Affecting
NOAEL and LOAED

Some assumptions on environmental variability have to be made to determine NOAEL and LOAED (i.e. cleanliness of the collection and a control of the environment, such as ensuring an average RH between 50 and 60% and a temperature range between 20 and 30°C). If these requirements are not met, NOAEL and LOAED values may need to be re-evaluated. The potential effects of the state of the material and abnormal or unacceptable levels of the most critical environmental parameters on the NOAEL and LOAED need to be assessed.

Relative Humidity

Water vapour has been selected as a key pollutant due to its significant impact. Even when water is not directly involved in the deterioration, it often has a major influence on adverse effects from other pollutants. Figure 22 shows a compilation of the relative LOAED shift of different pollutant–material systems at different RHs. The effect of RH seems rather specific for each system. The strongest effect of RH has been observed for NO_2–curcumin (a red pigment) system for RHs above 50% and for a hydrogen peroxide – black-and-white photograph system at RHs below 50%. Some systems show no significant shift. The LOAED of about half of the pollutant–material systems compiled has been reduced by half with a 10% increase in RH.

Temperature

In 1889, Arrhenius developed a mathematical relationship between reaction rate and temperature (Arrhenius 1889). This relationship has often been used to predict the behaviour of organic materials over time. Sufficient experiments on thermal degradation were done to confirm the Arrhenius relationship for the degradation of paper by hydrolysis. The deterioration rate decreased by about 50% for each 5°C reduction at a constant RH (Wilson 1995). Some cyan and yellow colour dyes on photographs and cellulose acetate films show similar trends, as shown in Figure 23. The adverse effects observed after a few decades at room temperature will take more than 100 yrs at 10°C.

It is often more convenient to combine RH and temperature parameters when attempting to predict the deterioration of organic materials. Figure 24

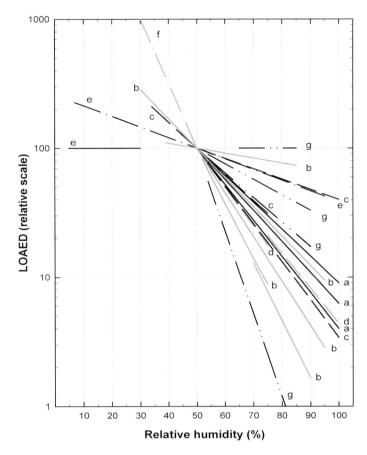

Figure 22. Effect of RH on LOAED for: (a) acetic acid or formic acid on copper; (b) COS, H_2S, or SO_2 on copper; (c) acetic acid or formic acid on lead; (d) acetic acid on tarnished lead; (e) COS or H_2S on silver; (f) hydrogen peroxide on a silver image; and (g) NO_2 on colorants. All measurements done at 20–25°C. Data from Appendix 2.

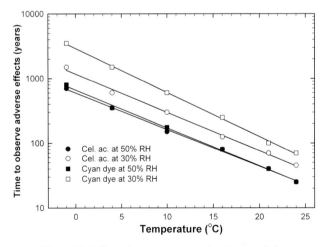

Figure 23. Effect of temperature on the time needed to observe adverse effects on cellulose acetate films and colour dyes in photographs at 30 and 50% RH. Data from the Image Permanence Institute, Rochester, New York. The critical adverse effects chosen were 0.5 free acidity for cellulose acetate films and 30% (dark) fading of the least stable dyes (cyan) for early 1990s colour slides, negatives, and prints (Reilly 1993, 1998).

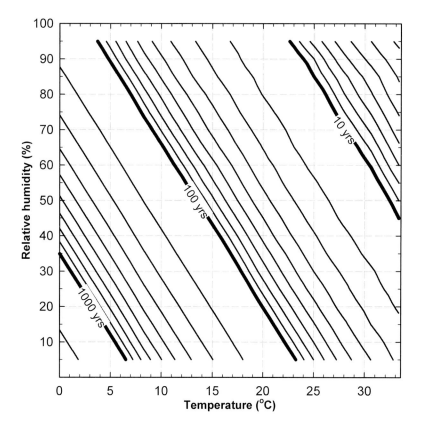

Figure 24. Prediction of observable adverse effects on materials sensitive to hydrolysis at various combinations of RH and temperature. Data based on accelerated ageing of cellulose acetate film.

shows the time required for some organic materials sensitive to hydrolysis (e.g. magnetic recording tape, colour prints, and flexible PVC) to suffer observable adverse effects at various combinations of RH and temperature (see Table 2 for additional moisture-sensitive materials). The prediction of observable adverse effect is based on cellulose acetate film, which is quite sensitive to hydrolysis [activation energy: 92 kJ/mol; adverse effects are noticed after 44 yrs at 20°C and 50% RH according to Reilly et al. (1995)]. Figure 24 can be useful in determining the most cost-effective RH and temperature set-up for preservation.

RADIATION

Radiation, especially UV and visible, can initiate and speed up some deterioration processes. Photo-oxidation is a good example of the action of radiation on materials. Photo-oxidation can happen in the presence of light ($h\nu$) and water vapour and/or oxygen. When a molecule of the material (M) is photo-excited (M*), it can undergo various photochemical reactions, as simplified

below. It can react with water and hydrogen peroxide (HOO·) formed in a second step from oxygen, and the hydrogen radical (H·) can destroy the molecule. It can also react with oxygen leading to singlet oxygen, a highly reactive excited state of the oxygen molecule (O_2^*), which can also destroy the molecule directly or indirectly.

$$M + h\nu \rightarrow M^*$$

$$M^* + H_2O \rightarrow \cdot MOH + H\cdot$$
$$H\cdot + O_2 \rightarrow HOO\cdot$$

$$M^* + O_2 \rightarrow M + O_2^*$$
$$O_2^* + H_2O \rightarrow HO\cdot + HOO\cdot$$

DIRT ON OBJECTS

Even without supportive data, it is safe to assume that depositions of salt, oils, metal particles, and cleaning product residues have a significant affect on the NOAEL and LOAED. In many cases, corrosion begins when a particle is deposited on a metal surface. Salts are hygroscopic and are active in corrosion or efflorescence processes. Salt-contaminated seashells can probably have a NOAEL of acetic acid vapour that is 3–100 times lower than that for "clean" shells (Figure 25). Decontamination of shells and low fire ceramics can eventually become an important specification for their preservation. Oily residues tend to absorb airborne pollutants, and metal particles can behave as catalysts* in the deterioration process.

MIXTURES OF POLLUTANTS

The possible adverse effects of gas mixtures on the NOAEL and LOAED for different airborne pollutant – object systems can be found in the literature. Though experiments were often made at abnormally high levels, synergistic effects* on the LOAED were not generalized apart from the water vapour (RH) as shown above. The opposite effect is actually observed in a few experiments. For example, the corrosion rates of lead by acetic acid decreased with the addition of formic acid (Tétreault et al. 2003). No change was observed on lead and copper when formaldehyde was added to the formic acid rich environment. However, a high concentration of pollutants can influence the rate of deterioration and, in some cases, the nature of the alteration layer on the object's surface without a significative impact on the NOAEL and LOAED. A compiled list of the

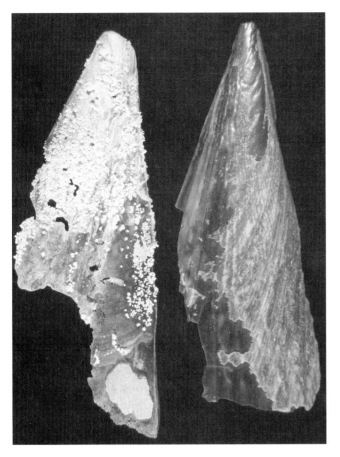

Figure 25. Efflorescence on seashells. The shell on the left has been exposed to acetic acid vapour and high RH. The shell on the right is the control sample. This type of deterioration is also known as Byne's disease in honour of Mr. Loftus St. George Byne who was the first to describe it in 1899 (Byne 1899). This type of efflorescence should not occur as long as RH is controlled, the sample is kept clean, and there are no materials nearby that emit high levels of acids. [A colour version of Figure 25 is available on p. 94.]

influence of gas mixtures on the first signs of adverse effects on objects is presented in Appendix 3.

State of Materials

Other parameters that can influence the NOAEL and LOAED include the actual state of the materials and their level of protection. Already faded artists' colorants have a higher resistance to pollutants (higher NOAEL or LOAED) than do new colorants. The LOAED for a half-faded colorant will shift up by about 1 order of magnitude. This is due to the auto-retardant pattern where the colour change is faster in the early stages of fading than in the final stage. The same magnitude of shift happens for half-faded colorants under visible light. In a similar way, silver will deteriorate more slowly (slower formation of

silver sulphide) if partly tarnished. Unfortunately, the opposite scenario has also been observed for other objects which deteriorate in an autocatalytic way when the formation of new products speeds up the deterioration process. This is the case for cellulose acetate- and nitrate-based films and acid papers in airtight enclosures.

Most of the characterizations of pollutant–material systems reported in Table 3 are based on new materials. In reality, objects often have been exposed to an uncontrolled environment before acquisition by a museum. Once they become part of the collection, the environmental control offered by the museum does not necessarily provide adequate control of pollutants or other agents of deterioration. The past exposure of many books to high levels of sulphur dioxide must be kept in mind when extrapolating their future preservation condition.

An induction period for some pollutant–material systems has been noticed where, even in the presence of a substantial amount of pollutants, no adverse effect was noticed for periods varying from a few months to a few years. Most documented induction phenomena have been observed related to the hydrolysis of organic objects, particularly with cellulose acetate- and nitrate-based films containing plasticizers, stabilizers (such as antioxidants in polyethylene), and other possible additives which have given temporary protection. This induction can falsify the determination of the NOAEL or LOAED.

Stresses or static forces on objects can provoke deterioration at lower NOAEL or LOAED. This has been well observed with stress vulcanized natural rubbers, such as inflated automobile tires.

Barriers and Binders

In some cases, a vulnerable object is protected from pollutants with a barrier material (binder or film). A colorant can be mixed with a resin (binder). The thicker the binder, the more difficult it is for the pollutant to diffuse itself through it. Only the very top pigment layer will fade and the under layers will be much less affected. For this reason, watercolours are more sensitive to fading than oil paints. Any film at the surface of a colorant acts as a barrier against pollutants and other agents of deterioration. This can be, for instance, an oil or acrylic varnish on the surface of a painting or a lacquer applied on silver. The effect of a protective film or binder has been quantified with the fading action of ozone on the colorant alizarin crimson (see Table 3).

RISK ASSESSMENT 3

To determine the risk of deterioration for a particular object in a specific environment, it is necessary to identify the nature of each material in that object as well as the level of each possible harmful pollutant in the surrounding environment. This type of assessment, called "micro-scale" risk assessment (object level), can be quite time consuming for a mixed collection. Although a few homogeneous objects can easily be characterized in detail, a heterogeneous collection requires much more work. When the material science knowledge is limited or when the types of objects are varied, a "macro-scale" risk assessment (collection or room level) should be considered. A conservator can provide valuable help for both micro- and macro-scale risk assessment.

MICRO-SCALE RISK ASSESSMENT

The potential risk to a specific object can be evaluated by comparing the NOAEL or LOAED of airborne pollutants for each relevant component of the object with the pollutant levels in the surrounding environment. A list of typical levels of pollutants generated by different sources in enclosures or in rooms has been summarized in Table 4. Outdoor levels have also been included as a reference. These levels represent current knowledge. They may not reflect the reality of a specific museum, but they can serve as a preliminary step before starting an exhaustive monitoring program. A more extensive list can be found in Appendix 1.

Example 1 shows how the assessment can be done for the preservation of lead and copper in acetic acid environments. The NOAEL and LOAED of acetic acid for both metals have been juxtaposed with the probable levels of acetic acid generated by different sources. As can be seen in the example, acetic acid at levels higher than 400 μg m^{-3} can be harmful for lead objects. This means there is a good probability that lead will corrode in wooden enclosures. In contrast, copper reacts with acetic acid at a LOAED of 1000 μg m^{-3} yr, and only the most acidic wood species, fresh paint, silicone caulking, or cellulose acetate films with a strong odour will cause adverse effects to copper in enclosures over the short term. By excluding these products, there is little risk of short-term deterioration of copper.

When an object is composed of more than one material, the risk assessment should be made on the most sensitive pollutant–material system or in relation to a reference material, such as a moderately sensitive material. When many objects have to be assessed, this can become a very demanding task. In some cases it may be preferable to conduct the risk assessment on a macro scale as discussed below.

MACRO-SCALE RISK ASSESSMENT

It may be feasible to consider a set of objects or the overall collection in a specific location as having a similar vulnerability to pollutants. Although

EXAMPLE 1.

VULNERABILITY OF LEAD AND COPPER IN VARIOUS ENVIRONMENTS (DATA FROM TABLES 3 AND 4)

Level	0.1	1	10	100	1000	10 000	100 000
Acetic acid μg m^{-3}		0.3 ————————— 30 outdoor clean / polluted sites					
			rooms 40–100 oak 300————————7000				
			wooden enclosures 80 ——————— 2000				
				oil paint dried 5 weeks 20 000 —70 000			
		emulsion or two-part epoxy paint dried 5 weeks 3000 ——— 20 000					
		acid-type silicone cured 7 and 29 days 100 ————————1000					
		smelly cellulose acetate film collection 900 ————————————100 000					
Lead				NOAEL: 400 μg m^{-3}			
Copper		LOAED: 1000 μg m^{-3} yr: 10 μg m^{-3} 100 μg m^{-3} 1000 μg m^{-3}					
		for 100 yrs for 10 yrs for 1 yr					

Table 4. Levels of airborne pollutants[a]

Level (μg m^{-3})	0.1	1	10	100	1000	10 000

Acetic acid
0.3 ———————————— 30 outdoor clean/polluted sites
rooms 40 ——100   oak 300 ————————— 7000
wooden enclosures 80 ————————— 3000
oil paint dried for 5 weeks 20 000 – 70 000
emulsion or two-part epoxy paint dried for 5 weeks 3000 ——- 20 000
acid-type silicone cured for 7 and 29 days 100 ——————1000
smelly cellulose acetate film collection 900 ————————— 100 000

Ammonia
0.7 ———————— 20 outdoor clean/polluted sites
rooms 0.6 ————————— 60
400 to 3000 visitors per day in a building 10 ———- 30

Carbonyl sulphide
0.7 — 1 outdoor clean/polluted sites
0.03————————-1 wool in a room (dark and sunlight)

Formaldehyde
0.5 ———————— 30 outdoor clean/polluted sites
rooms 10 ————— 70
wooden enclosures 50 ————— 500
wood products with urea formaldehyde 500 ————— 6000

Formic acid
0.1 —————————————20 outdoor clean/polluted sites
0.1 ———————— 30 rooms
wooden enclosures 2 ————————- 2000

Hydrogen sulphide
<- 0.01 ————————— 10 outdoor clean/polluted sites
0.03————————————- 40 rooms (including visitors)

Nitric acid
0.4 ———————— 30 outdoor clean/polluted sites
rooms 2 ————— 20

Nitrogen dioxide
outdoor clean/polluted sites 2 ————— 100
rooms 2 ————— 90
cellulose nitrate negative films 4000

Ozone
outdoor clean/polluted sites 2 ————— 400
rooms 0.1 ————— 100
rooms with ozone generator 4 ————— 600
or photocopy machines

Particles, coarse (10 μm)
outdoor clean/polluted sites 2 —————70
rooms 1 ————— 100

Particles, fine (2.5 μm)
outdoor clean/polluted sites 1 ————— 50
rooms 1 —————-30

Sulphur dioxide
0.1 ————————- 100 outdoor clean/polluted sites
rooms 0.1 ————— 50

	0.1	1	10	100	1000	10 000

a: Details of the sources and monitoring conditions are in Appendix 1. If not specified, the levels of pollutants are generated by the material inside airtight enclosures.

convenient, this approach must be used cautiously. Table 5 addresses this issue. It shows the maximum allowable concentration for each key airborne pollutant to provide minimal risk to most materials for the indicated exposure periods. The targets are based on the LOAED of pollutants. These maximum levels of pollutants have been grouped in three preservation targets*: 1, 10, and 100 yrs. The 100-yr target provides the greatest preservation and means that most objects should not show adverse effects for 100 yrs when exposed to the maximum level of pollutants allowed in the respective column. As already mentioned, with the exception of water vapour, it is probably unrealistic and unreliable to specify pollutant concentrations lower than those allowed by the preservation target for 100 yrs. Intermediate targets can also be adopted. Another way to visualize preservation targets is to refer to the rate of observable adverse effects (number of adverse effects per unit time), as shown in Figure 26. A preservation target of 1 yr has an observed adverse effect after 1 yr. With each successive year additional adverse effects will be observed. For a preservation target of 10 yrs, the adverse effects will be seen after 10 yrs, and will be cumulative with each successive 10-yr period.

It is important to stress that hypersensitive materials (e.g. lead, silver, vulcanized natural rubber,

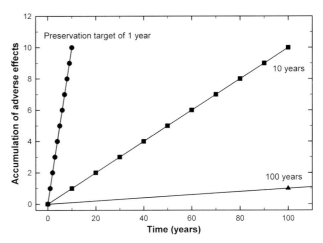

Figure 26. Rate of accumulation of adverse effects for preservation targets of 1, 10, and 100 yrs.

polyurethane magnetic tapes, cellulose acetate- and nitrate-based objects, and some colorants) have not been included in Table 5. These are so sensitive that they should be dealt with individually, with proper control strategies. Details of their preservation are addressed in Table 15 (p. 62). By excluding the materials that are hypersensitive to each key pollutant, the LOAED of a pollutant for a mixed collection is shifted up by an order of magnitude. In other words, the preservation targets are based

TABLE 5. AIR QUALITY TARGETS FOR MUSEUM, GALLERY, LIBRARY, AND ARCHIVAL COLLECTIONS

Key airborne pollutants	Maximum average concentrations for indicated preservation targets[a], $\mu g\ m^{-3}$ (ppb)			Reference average concentration range, $\mu g\ m^{-3}$	
	1 yr	10 yrs	100 yrs	Clean low troposphere	Urban area
Acetic acid	1000 (400)	100	100[b]	0.3–5	0.5–20[c]
Hydrogen sulphide	1 (0.71)	0.1	0.01	0.01–1	0.02–1
Nitrogen dioxide	10 (5.2)	1	0.1	0.2–20	3–200
Ozone	10 (5.0)	1	0.1	2–200	20–300
Sulphur dioxide	10 (3.8)	1	0.1	0.1–30	6–100
Fine particles ($PM_{2.5}$)	10	1	0.1	1–30	1–100
Water vapour	keep below 60% RH[d]				

a: Preservation target is the length of time (in years) for which the objects can be exposed to the indicated level of pollutants with minimal risk of deterioration. These targets are based on the LOAED for most objects and assume that average RH is kept between 50 and 60%, temperature ranges between 20 and 30°C, and the collection is kept clean (if not, the maximum levels of key airborne pollutants for each class of targets may need to be readjusted). These values are not applicable to hypersensitive materials.

b: Because most objects have high NOAEL of acetic acid, concentrations below 100 $\mu g\ m^{-3}$ are not mandatory.

c: Acetic acid levels can be as high as 10 000 $\mu g\ m^{-3}$ inside enclosures made with inappropriate materials, such as fresh acid-cured silicone.

d: For permanent collections where the average RH has not been between 50 and 60%, maintain the historical conditions.

on moderately sensitive objects and not on those that are the most sensitive. This rule does not apply to sulphur dioxide because paper-based materials are very sensitive to SO_2 but also very common. Another exception is the LOAED approach for water vapour. This can hardly be applied to a mixed collection because, although reducing the humidity levels can be beneficial for materials prone to hydrolysis, many composite objects can be physically damaged by changes in RH.

The maximum concentration of key airborne pollutants allowed in Table 5 for the preservation targets of 1 and 10 yrs should be considered effective and relevant for preservation purposes. Many objects will not show the first signs of deterioration until after the designated exposure period. The targets of 1 and 10 yrs are useful and feasible for many museums and, in fact, the best achievable in most historical buildings. Achieving pollutant levels associated with a preservation target of 1 yr in a room can significantly contribute to better preservation in an enclosure if there are no significant emissive sources of pollutants inside.

Monitoring technology to measure all the key pollutants at levels as low as those required for the preservation target of 100 yrs is available (details of monitoring techniques are given in Chapter 6). However, most sensitive techniques require expensive instruments or sampling and laboratory analysis.

Not all museums can justify the cost related to monitoring. Fortunately, it is possible to provide an estimation of preservation targets based on some assumptions. Table 6 shows possible preservation targets for most collections in different museum locations, including rooms and enclosures. Parameters, such as public access, filtration and barrier capacity of the building, and emissive and sorbent materials in enclosures, have been considered for the establishment of these possible targets. The museum location is assumed to be in a moderately polluted urban area. The preservation targets have been assessed by comparing the maximum pollutant levels allowed for each preservation target in Table 5 with the data in Appendix 4, which reviews actual knowledge about the possible levels and reduction ratios of pollutant levels by different building features and enclosure designs. These potential targets may not reflect the reality of a specific museum; however, they can be used as a starting point for monitoring as well as a good reference for institutions which cannot afford a comprehensive monitoring campaign. If the resources are available, some monitoring may help to improve the estimation of the potential preservation.

The NOAEL and LOAED approaches can become sources of rational decision-making for the development of the institutional preservation policy*. The use of micro- and macro-scale risk assessment is dealt with further in Chapter 5.

TABLE 6. POTENTIAL PRESERVATION TARGETS FOR MOST COLLECTIONS

| Air quality control in building | In rooms | Potential preservation target (in yrs)[a] | | | |
| | | In enclosures with EM[b] | | In enclosures without EM | |
		Without ES[c]	With ES	Without ES	With ES
Natural ventilation or HVAC system with moderate-efficiency particle filter, no gas filter	1–10	≤1	10–100	10–100	≥100
HVAC system with gas and good-efficiency particle filters[d], building membranes that are good gas barriers, and basic control of visitor flow	10–100	≤10	10–100	≥100	≥100
HVAC system with gas and high-efficiency particle filters, building membranes that are good to very good gas barriers, and limited access	≥100	≤10	10–100	≥100	≥100

a: Adverse effects of water vapour and hypersensitive materials are excluded.
b: Emissive materials (products and objects).
c: Efficient sorbent (enclosures are assumed to have an air exchange rate of once per day).
d: Assumes periodic replacement of the filters.

CONTROL STRATEGIES 4

Control strategies are co-ordinated measures to reduce or maintain one or many agents of deterioration at a certain level, thereby limiting the risk or the rate of deterioration to objects exposed to the harmful agents. These measures can be derived from specifications (which are accurate descriptions of technical requirements for the performance of building features, portable fittings, and procedures). Table 7 summarizes the various control strategies that can be used to prevent adverse effects of airborne pollutants. It shows the possibilities for control at different stages. Avoid, block, dilute, and filter/sorb are strategies to reduce the levels of the pollutants in the ambient air; reduce reaction is a strategy to minimize the adverse effects of the pollutants on objects; and reduce exposure is a strategy to limit the deterioration by limiting the exposure of objects to the harmful environment. Whenever feasible, avoiding sources of pollutants is the best option. However, there are few options available to avoid outdoor pollutants, the most realistic being the blocking strategy. For indoor-generated pollutants, the avoid strategy (i.e. avoiding exposure by selecting safe products) is the most efficient choice for enclosures. If this cannot be done, the dilution strategy will provide a partial reduction in pollutants. Monitoring is an integral part of the control strategy, and is discussed in Chapter 6.

It is important to note that control strategies do not have to be uniformly applied throughout the museum. They can be tailored to the value and need for preservation of the objects as well as to different rooms and different enclosures; see Table 8 for examples of the many possible locations of objects (such as on display, in storage, in use, etc.). Control strategies are not intended to offer solutions to specific problems, as there are too many possible scenarios. Chapter 5 provides tools for selecting control strategies that meet the required performance. The current chapter proposes general guidelines followed by a focus on environmental control at the building, room, and enclosure levels. Some specifically adapted control measures are proposed for the preservation of hypersensitive objects. However, before discussing the possible control strategies, some general principles and guidelines are described to help explain the global physics and chemistry of pollutants in the museum.

GENERAL PRINCIPLES AND GUIDELINES

SPATIAL GRADIENT AND STEADY-STATE CONCENTRATIONS

Airborne pollutants can be emitted by some materials and sorbed by others. Figure 27 (p. 38) illustrates the mass transfer of airborne pollutants in ambient air including infiltration and exfiltration through different envelopes and from emission and sorption by various materials. All these parameters induce spatial gradients* in a pollutant's concentration in the room and provide a steady-state concentration* in a small enclosure. For simplicity, the influence of air movement (draft or pressure gradients) and thermal gradients are excluded in this schema. Because museums have natural ventilation due to periodic opening of doors or windows, the concentration of pollutants is reduced as they infiltrate deeper into the building (Blades et al. 2000). Pollutants get sorbed along their pathways by the surface of the walls, floors, and furniture as well as by the collection itself. A similar pattern can be seen in enclosures. Pollutants can infiltrate an enclosure through gaps or by diffusion through the envelope. Only a small percentage of the ambient pollutants will succeed, and these will be sorbed by the interior surfaces of the enclosure's products and by the objects. Only those sorbed by the object have the potential to cause adverse effects.

"100, 10, 1" RULE FOR THE LEVEL OF OUTDOOR POLLUTANTS

Based on a review of the literature and a model of the levels of pollutants found outside and inside enclosures, the "100, 10, 1" rule of thumb can be used to approximate the levels of outdoor pollutants passing through successive protection envelopes. Through each envelope, the concentrations of pollutants are reduced by 1 order of magnitude. For example, if there are 200 μg m^{-3} of nitrogen dioxide outside, the level inside a room in a building will be 20 μg m^{-3} and the level inside an enclosure in the room will be 2 μg m^{-3}. This assumes the absence of indoor-generated pollutants and the presence of good building barrier controls, collections in the rooms, and good airtight enclosures. If a room has windows that are left open a few hours a day, there will be no significant difference between levels on

TABLE 7. CONTROL STRATEGIES TO PREVENT THE ADVERSE EFFECTS OF AIRBORNE POLLUTANTS

AVOID outdoor sources
- Select proper locations for new buildings based on surrounding sources of airborne pollutants such as pollution-emitting industries and dominant winds.
- Minimize the generation of pollutants, e.g. pave the parking lot, limit traffic in the immediate vicinity of the building.
- Support any proposal for reduced coal energy consumption and support environmentally friendly initiatives.

AVOID sources in rooms/building
- Minimize dust- and gas-generating activities close to the collection or in the same ventilation zone.
- Limit the number of visitors per room depending of the ventilation capacity.
- Carefully select and use products based on their chemical components.

AVOID sources in enclosures
- Carefully select and use products based on their chemical components.

BLOCK infiltration of pollutants in rooms/building
- Consider wise distribution of collections in the different air quality zones of the building (including the possibility of using enclosures).
- Improve airtightness of the building membrane or some rooms.
- Add vestibules for the main doors and open windows wisely.
- Select proper location for the air intake of the HVAC system, provide different positive pressure zones with a minimum air intake ratio. Insert efficient gas and particle filters. Change filters periodically.

BLOCK infiltration of pollutants in enclosures
- Use airtight enclosures or use air-filtered positive pressure systems. Change filters periodically.
- Wrap objects with sorbent tissues, such as acid-free tissues or cotton fabrics.

BLOCK emission of pollutants from products in rooms/enclosures
- Apply a barrier film on the surface of emissive wood products.

BLOCK transfer (deposition or sorption) of pollutants on objects
- Apply a barrier film on the surface of objects (limited option).

DILUTE, FILTER, or SORB pollutants in rooms/building
- Consider wise distribution of collections in the different air quality zones of the building (including the possibility of using an enclosure).
- Increase the distance between objects and the source of the airborne pollutant or its point of infiltration.
- Use local exhaust for the most polluting activities (cooking, workshop, chemical store).
- Use portable fans to push the air out of the rooms/building (for short-term high emission such as freshly painted walls or floors).
- Filter the recirculating air of the HVAC system or use a portable filter unit. Change filters as recommended by the manufacturer.

DILUTE or SORB pollutants in enclosures
- Consider stack pressure design of the enclosure if the environment of the room is well controlled or use air-filtered positive pressure systems.
- Dilute (flush) air with a non-reactive gas such as argon, helium, or nitrogen.
- Use passive or active sorbent methods (can include gas, particles, water vapour, or oxygen sorbent). Change the sorbents periodically.

REDUCE REACTIONS on objects
- Decrease the level of RH, the temperature, or the ultraviolet, visible, and infrared radiation (where appropriate).
- Neutralize pollutants absorbed into the objects (e.g. alkaline compounds in papers inhibit acid deterioration).

REDUCE EXPOSURE TIME
- Limit exposure of objects to the inappropriate environment.

MONITOR the collection
- Inspect for signs of deterioration on objects and on the enclosure and building products periodically.

MONITOR pollutants in rooms and in enclosures
- Do appropriate in situ monitoring of pollutants.

MONITOR performance of control features
- Verify efficiency of the gas and dust filter systems periodically.
- Measure the leakage rate of the building and enclosures.

RESPOND to the detection of pollutants or damage on objects
- Protect objects from the harmful environment.
- Re-evaluate the avoid, block, and reduce strategies; consider cost–benefit analysis.
- Remove dust accumulation on objects, in the building, and on portable fitting surfaces periodically. Minimize resuspension of dust.

TREAT damaged objects
- Clean and treat objects, if necessary, to limit further deterioration.

TABLE 8. VARIOUS POSSIBLE LOCATIONS OF OBJECTS

Function (duration[a])	Room	Enclosure with typical products
Storage (long term)	Mixed or specific collection room (staff or expert access)[b] Cool room (10°C) (limited access) Cold room (-20°C) (limited access) Dry room (<40% RH) Attic and basement of historical houses, shed	Wood or metal cabinets Wood, cardboard, metal, or plastic boxes/containers Plastic bags Plastic, paper, cardboard pockets/envelopes Wood product transportation boxes Sealed glass jars (wet collection: ethanol or formaldehyde) Plaster (some untreated fossil specimens)
Exhibition (short to long term)	Exhibition room (possible location of receptions) "Storage-exhibition" room (public access) Public area (subway, shopping mall) Political and diplomatic offices Private residences, religious building	Display cases, glass frames: Window: glass or plastic Base: metal, wood, plastic Frame/boards: metal, wood, plastic, cardboard
In use (very short to short term)	Ceremonies and events (spiritual, cultural, etc.) Maintenance	
Conservation laboratory (very short to long term)	Short- to long-term presence of an object in treatment or in analysis	Wood, cardboard, or plastic boxes Clear plastic or paper envelope (encapsulation)
Consultation or preparation room (very short term)	Immediate or short-term use of objects	Wood, cardboard, or plastic boxes Clear plastic or paper envelope (encapsulation)
Transit (short term)	Loading bay Quarantine room	Metal, wood, or cardboard transportation boxes (inside: soft and rigid foams, plastic sheet, papers, etc.)
Non-collection	Cafeteria, workshop room, main hall (reception), auditorium, and offices	
Construction activities (very short to long term)	New building, new room (mainly for exhibition), or retrofitting (renovation)	Mainly new display cases or retrofitting (various fresh products including those that release solvents)
Outside (moderate to long term)	Outside of heritage institutions, public and private areas (storage or exhibition): Open structure, sheathed post and beam	Large display cases (glass, plastic, metal, wood products)

a: Duration: very short term = less than a day; short term = 1 day to a few weeks; moderate term = a few weeks to 1 yr; long term = more than 1 yr.

b: Some collections, such as organic ethnographic, taxidermal, and study skins, some minerals, wet collections (fluid preserved), and cellulose acetate and nitrate film collections, tend to release significant amounts of pollutants that could affect the objects themselves.

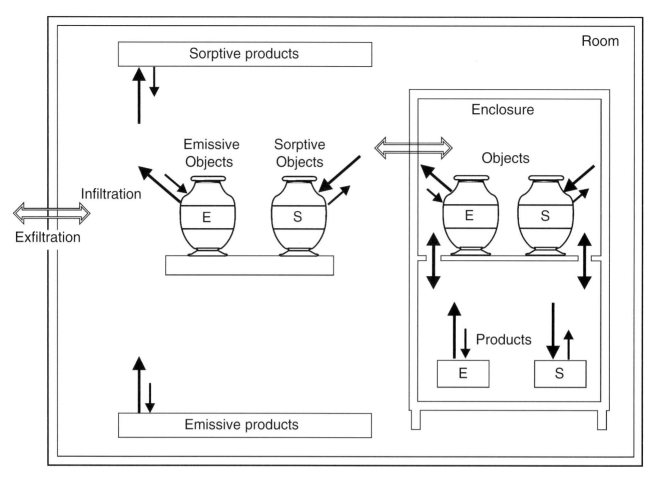

Figure 27. Schema of mass transfer of airborne pollutants in a museum. E = emissive; S = sorptive.

the inside and outside. The "100, 10, 1" rule for outdoor pollutants is a generalization and may seem simplistic, but it does satisfy the needs of many museums that cannot afford expensive monitoring but still require some idea of the order of magnitude of pollutant levels. Environmental departments of different levels of government disseminate (through their Web sites) the levels of pollution of major cities, and this information can be used to establish indoor levels using the "100, 10, 1" rule. See Chapter 6 for more information.

The "100, 10, 1" rule cannot be applied to indoor-generated pollutants such as acetic acid, which can reach high steady-state concentrations inside airtight display cases or storage cabinets made with emissive products. Research on formaldehyde in the 1970s and 1980s found levels in the range of 500–6000 $\mu g\ m^{-3}$ inside enclosures made from wood products with urea formaldehyde-based glue. Formaldehyde levels were 10 times lower in rooms with a large number of wooden panels (Meyer and Hermanns 1986; Newton et al. 1986), and would have been much lower still

if exfiltration of formaldehyde from an enclosure had been the only source of formaldehyde.

USEFUL LIFETIME OF PRODUCTS
Some products, especially those composed of organic materials, release pollutants in typical patterns. These patterns are related to their nature, the way the products are formed, and the deterioration processes. Based on each product's emission rate at different stages of its existence, the use of the products can be allowed under some conditions or can be prohibited completely. Figure 28 represents some possible patterns, based on products emitting acetic acid. Pollutant levels can be quite high in the early stages when products are made or being applied. This is the case for liquid products such as solvents or water-based paints and adhesives, glued wood products and in situ chemically made products such as room temperature vulcanization (RTV) silicones, and two-part products (urethane and epoxy) in the form of paint, adhesive, or moulding resins. It takes about 3–4 weeks for most of the off gassing to reach a minimal rate (Tétreault 1999a, 1992b). Much of the

damage reported in the literature as being caused by emissions from products concerned objects that were stored or displayed in newly made enclosures. Damage was noticed within the first 3 months.

After the initial period of curing or evaporation, some of these products release only a low steady level of pollutants due to the ongoing deterioration process of hydrolysis or oxidation. Many products are safe to use in enclosures in the "stable" phase. Most wood and plastic products that release acetic acid fall into this category, tending to create steady levels of pollutants that are sometimes below the detection limit. However, oak and cedar release very high amounts of acetic acid vapour. Their emissions are highest when freshly cut, and decrease slightly over time due to alterations to their surfaces (photo-oxidation, dust deposition). But even the reduced emission rates are significant, and a 50-yr-old piece of oak will still have a noticeable smell of vinegar. Because of the high levels of acetic acid generated by oak and cedar of any age, these woods are not appropriate for use in an enclosure that contains lead objects.

While some products are not recommended for use during their early stage of life, for others the scenario is the opposite. For example, some objects or products deteriorate after a period ranging from a few years to a few decades, and this deterioration causes an increase in the rate of emission of pollutants. Acidic papers and cellulose nitrate and acetate objects which suffer accelerating deterioration due to acid hydrolysis fall into this category.

It is important to consider the aspect of useful lifetime when choosing products, especially for enclosures. This may help to prevent the common adverse effects caused by various emissive products or objects at various points in their life span.

ADVERSE EFFECTS BY CONTACT

Another important aspect to consider when selecting products is their potential adverse effects when placed

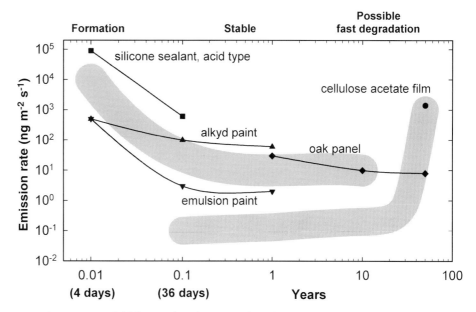

Figure 28. *Useful lifetime of products. Based on the rate of acetic acid emission from various materials (Tétreault 1999a; Ryhl-Svendsen 2000).*

in contact with objects. One potential problem is staining of an object at the contact point after a certain time. Therefore, it is important to ensure that no product can transfer any of its components to an object by contact. The products most likely to cause staining include products made of acidic components, such as acidic papers (Figure 29); paints formed by oxidative polymerization; poly(vinyl acetate) (PVAc) adhesives; dyes that are not colourfast; sulphur compounds; and products with high amounts of plasticizers such as flexible poly(vinyl chloride) (PVC). Polyurethane foams will also eventually cause staining by contact. They can maintain their initial properties for a few years but gradually turn yellow, brittle, and sticky. At this stage, they stain anything in contact with them. Contact between two objects or between components of the same object can also result in deterioration. The protection of products in contact with museum objects is also important. Acidic or oily objects can stain labels or their containers, which could lead to vital information related to the history or identification of the objects being lost. When a piece of metal is in contact with an acidic product, such as a wooden panel, most of the damage will occur at the edge of the metal rather than at the contact point. Moisture in the ambient air speeds up this deterioration.

An object can also damage its enclosure or container, which can lead to damage to other objects if the damaged product is re-used. For example, containers can be rusted or stained by

*Figure 29. Paper artwork stained by
an acidic matboard after many years of contact.
[A colour version of Figure 29 is available on p. 94.]*

oils from a mineral, ethnographic, or industrial object. If the container is used again for other objects, they may deteriorate due to contact with the contaminated container or by desorption of pollutants from it.

Adverse Consequences of Control Strategies

No control strategy should improve the preservation of one object at the expense of others, and no control strategy to reduce the risk of deterioration by pollutants should increase the risk of deterioration by other agents. These are two idealistic principles. In some circumstances, the selected control strategy is the lesser of two evils. Human health and safety should, of course, be mandatory.

Macro versus Micro Environmental Control

The decision of whether to exercise environmental control in an entire room or in selected enclosures depends on several parameters: the preservation needs of the objects, their sizes and quantities, aesthetic considerations, and resources. One common approach is to offer basic control at

the room level that can fulfil the institutional mandate of preservation and accessibility. Valuable objects at high risk can then be identified and receive extra control measures, such as displaying them in enclosures with a more highly controlled environment or limiting their display time. Museums with a large collection can group objects based on the degree of preservation desired, and then provide efficient control measures at the room level to meet these goals. A cost–benefit analysis can help with the decision to control the environment at the room or enclosure level. The issue of airtightness of enclosures is an important consideration and is covered below.

Conservation professionals can assist in grouping the collection based on sensitivity or common materials, and then apply specially adapted controls. Over time, some objects reaching an advanced state of deterioration may have to be relocated to avoid adverse effects on other objects. This is the case with objects made of cellulose acetate — they must be isolated from the rest of the collection when they reach a critical acidity to avoid cross-contamination.

Controls at Building/Room Level

The main issues at the building level are the infiltration of outdoor gaseous and particulate pollutants, and the emissions from indoor products, visitors, and, in some cases, the collection itself. Common or important control strategies are outlined in Table 7 and discussed below.

Avoid

There is usually very little that can be done to control the levels of pollutants outside buildings. However, some measures that can be applied locally include having paved parking spaces and a paved entrance way to avoid resuspension of dust, and limiting heavy traffic close to the museum. For indoor sources, there are many more possibilities.

Control of Visitor Density

In a crowded exhibit room with inadequate ventilation, the level of pollutants such as ammonia, hydrogen sulphide, dust, and water vapour may increase; likewise the temperature may rise. For the comfort of both visitors and the collection, limiting the maximum density of visitors per room is recommended. For popular exhibits, this can be done by allowing only a certain number of visitors to enter the exhibition every half hour.

TABLE 9. PARAMETERS OF HVAC SYSTEMS RELATED TO THE CONTROL OF POLLUTANTS

Parameters	Advantages	Limitations
Low air intake ratio	• Can be maintained to the minimum required for human health. • Avoids infiltration of outdoor pollutants and maintains the conditioned RH and temperature. The air intake location should be away from the street, exhaust vents, sources of pollution (chemical store, garbage containers), and the truck loading entrance.	• Important to avoid elevated levels of CO_2 [<1800 $\mu g\ m^{-3}$ (<1000 ppm) (ASHRAE 2001a)], so a CO_2 sensor is recommended. Typical background level of CO_2 is around 400 ppm, and it is increasing. Also must avoid elevated levels of indoor-generated pollutants [total VOC (TVOC) below 200 and 300 $\mu g\ m^{-3}$ during occupancy as suggested by Molhave (1990) and Seifert (1990), respectively.] • Sufficient amounts of outdoor air are required to provide positive pressure. This avoids the undesirable introduction of untreated air from open doors and leak points. These leak points should be sealed as much as possible. • High ratios of outside air to remove (flush out) VOCs from new products are not recommended if the collection is present.
No air economizer (no relief air)	• Avoids introduction of a large amount of untreated outdoor air.	
Multi-environmental control zones	• Provide different performances adapted to the needs of the collection. The highest positive pressure is usually applied to the cleanest room.	

Variable air volume (VAV)	• Airflow can be adjusted to the need: an increase of flow with a high number of visitors or renovation activities and a reduction during the low occupancy period. The option of having constant air volume is possible.
	• The following airflow rates are proposed as guidelines (Thomson 1986):
	  ceiling height <3 m: 8 air exchanges/h
	  ceiling height 4–5 m: 6 air exchanges/h
	  ceiling height 7 m: 4 air exchanges/h
	  These airflows can be reduced by 50% during non-occupancy.
		• A minimum flow is required to avoid gradients of temperature and RH and to control indoor-generated pollutants and avoid elevated levels of CO_2
		• The efficiency of the gas sorbents decreases when the airflow increases; noise will also increase.
		• ASHRAE has concerns related to efficiency and reliability of the VAV mode and recommends the constant air volume mode (ASHRAE 2003). It is also more difficult to maintain pressure relationships between adjacent HVAC zones with VAV than it is with constant systems.

Use of high-performance particle filters	• Allows better filtration of fine particles. See Table 10. • Delays the need for cleaning HVAC components.	• Performance can be lower than predicted if inadequately installed. • Indoor-generated dense particles are not trapped by the HVAC system due to their short suspension times. • A more powerful pressure fan may be needed for higher efficiency filters (more energy consuming). • Electrostatic air cleaners cannot be used because they generate ozone.
Use of gaseous and vapour sorbents	• Allows filtration of gaseous and vapour pollutants (may need a combined chemical sorbents filter system). See Table 11. For new construction projects, consider leaving space for future installation of a gas filter box or bag even if it is not currently requested by the client.	• Performance can be lower than predicted if inadequately installed. • Lifetime determination of some filters can be done by sampling analysis. If lifetime is unknown, the filter should be replaced at least once per year. Some released sorbent particles are trapped by the fine filter downstream while a very small amount pass through. Some of these particles can oxidize material surfaces. Fixed-bed filters do not tend to release sorbent particles. • Some filters require a more powerful pressure fan, which is more energy consuming.
Filtration of the return air	• In addition to the filtration of outdoor pollutants, indoor-generated pollutants such as H_2S, ammonia, and particles emitted or shed from visitors will be filtered.	• More energy consuming.
Regular inspection and maintenance	• By conducting regular inspection and maintenance, it is possible to ensure proper efficiency and to minimize the risk of microbiological growth. Newly installed HVAC systems should be cleaned before being operated. HVAC systems often become dirty during construction activities within a building, and should be cleaned. The air handling unit should be inspected by visual means every year and the supply and return ductwork every 2 yrs. Before beginning cleaning work, staff conservators should be informed of the protocol. • Do not use acid or oxidant solutions to clean cooling coils and ducts. • Avoid vegetable oil-based lubricants and keep the oil residues on HVAC components to a minimum. • Dusty filters are known to be more efficient at reducing particle infiltration than clean filters. However, dusty filters cause the pressure of the system to drop and the dust accumulated on the filter can be a source of pollutants and moulds.	

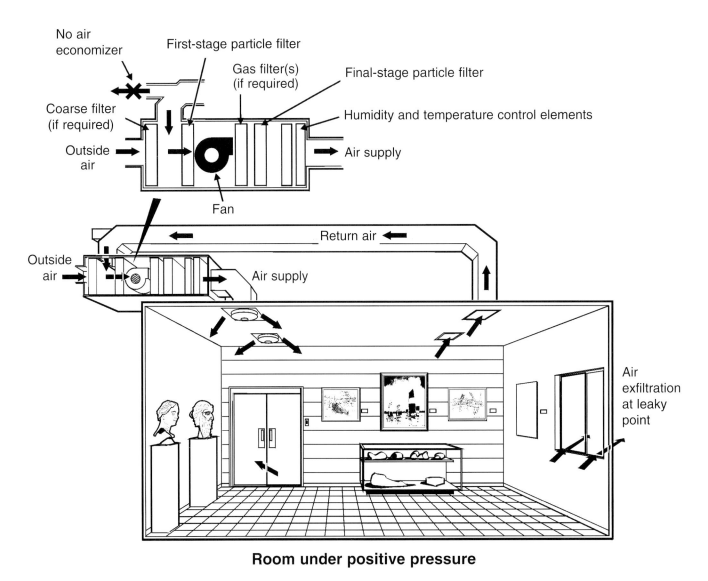

No air economizer

First-stage particle filter

Gas filter(s) (if required)

Final-stage particle filter

Coarse filter (if required)

Humidity and temperature control elements

Outside air

Air supply

Fan

Return air

Outside air

Air supply

Air exfiltration at leaky point

Room under positive pressure

Figure 30. Typical HVAC system.

Selection and Use of Products

Due to the high dilution capacity of a room (from natural or forced ventilation), pollutants generated by products do not tend to reach high levels beyond a few weeks following their application. However, three products should be avoided in large quantities: oil or alkyd-based paints and varnishes (see Table 13, p. 53); wool carpet; and uncoated wood products such as particleboard or waferboard made with urea formaldehyde-based glue. Noticeable smells in a room are often related to the collection itself on open shelves or to moisture problems. If there are silver and copper objects in a room with wool tapestries or carpets, these metals should be protected. There is usually no need to replace old oil-based paints or old uncoated wood panels if there is not a pronounced smell.

However, as a precaution, objects having lead components should be protected.

BLOCK, DILUTE, FILTER, OR SORB
Control with an HVAC System[2]

Large new buildings control the level of pollutants through a central HVAC system especially designed for this purpose. Such systems must fulfil requirements for human health and for the preservation of the collection at a minimum cost. Basic HVAC systems tend to concentrate on a stable

2. Since 1999, the manual of the American Society of Heating, Refrigeration, and Air-Conditioning Engineers (ASHRAE) has had a chapter dedicated to museums and libraries which focusses mainly on RH and temperature issues. In the 2003 version, controlling airborne pollutants has been added.

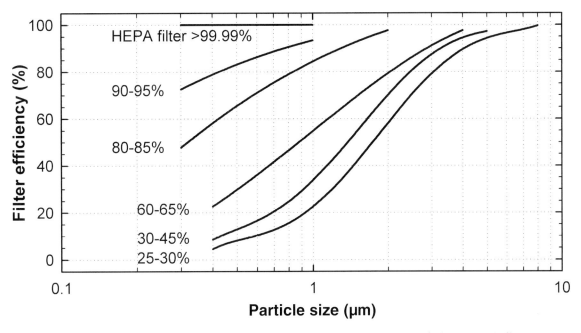

Figure 31. Efficiency of filters versus particle size based on the dust spot test (ASHRAE 2001d).

and uniform climate with modest control of dust. In historic houses, the climate control performance depends not only on the HVAC system but sometimes even more on the vapour/gas barrier and thermal isolation capacities of the building and on human activities. Accurate control of pollutants requires expert assessments to determine the combined performance of the HVAC system and the building envelope.

Many parameters of the HVAC system are involved in the control of pollutants. Table 9 (p. 41) summarizes these parameters and Figure 30 gives a visualization of them. Variants of this schematic system exist. The air intake should be the lowest possible that will still be sufficient to keep levels of CO_2 and indoor pollutants low by dilution and to pressurize the different control zones of the building. Because weather and indoor activities are not constant and the cost of energy is increasing, airflows based on demand (variable air volume system) have gained in popularity. Although these systems are satisfactory in many applications, they may not be suitable for museums because the low and variable airflows could result in temperature gradients that in turn create incorrect RH during warm or cold days, and because they may be incapable of reaching the required environmental performance (ASHRAE 2003). When a variable air volume system is not the best option, museums should consider having different environmental control zones optimized for the preservation of different types of collections and

for the functions of individual rooms. ASHRAE refers to different methods for the measurement of the airflow rate of rooms (ASHRAE 2001b).

Filter systems are important in the control of pollution in new or retrofit buildings. An HVAC filter system can have different filter configurations ranging from a simple water spray with a coarse particle filter to a complex series of specialized gas and high-performance particulate filters. ASHRAE and the European Committee for Standardisation (CEN) have developed new performance standards for dust filters: the ANSI/ASHRAE Standard 52.2 (ASHRAE 1999a) and the EN 779 (CEN 2002). Unfortunately, there is no direct equivalency between these two standards. The previous standard dust spot test and the (old) EN 779 (ASHRAE 2001d) are based on the same measurement method and can be used to make approximate correlations between the two new standards. Figure 31 shows the typical efficiency of filters for different particle sizes based on the dust spot test. These efficiencies serve as estimations, since the real efficiency varies depending on such physical parameters as airflow and saturation of the filter. For example, a filter with an efficiency of 80% blocks half of the 0.3-μm particles and more than 95% of the 2.5-μm particles. As shown in Figure 7, blocking infiltration of fine particles is important since the most harmful particles for museums are those having an aerodynamic diameter between 0.01 and 1 μm, such as carbon black, salt, and sulphate and nitrate compounds.

TABLE 10. SPECIFICATIONS FOR HVAC FILTER SYSTEMS

Class of specification	First-stage particle filter, minimum efficiency based on			Gas filter[d]	Final-stage particle filter, minimum efficiency based on			Return air[e]
	ASHRAE 52.2 (MERV[a])	Dust spot efficiency[b]	EN 779[c]		ASHRAE 52.2 (MERV)	Dust spot efficiency	EN 779	
AA	≥12	≥70%	≥F6	1 or 2 stages	≥16	>99%	≥H10	filtered
A	11	60–65%	F6	1 stage	15	>95%	F9	filtered
B	10	50–55%	F5	1 stage (preferable)	14	90–95%	F8	filtered
C	9	40–45%	F5	none	13	80–85%	F7	filtered (preferable)
D	8	30–35%	G4	none	12	70–75%	F6	unfiltered

a: MERV: Minimum efficiency reporting value from ANSI/ASHRAE Standard 52.2 (ASHRAE 1999a).
b: Performance of the filter based on the atmospheric dust spot test ASHRAE 52.1 (included for reference) (ASHRAE 1992).
c: EN 779 from the CEN (CEN 2002).
d: See Table 11 for the selection of gas filters.
e: Return air is filtered when it is recirculated through the filter system.

The performance of the filter system should be proportional to the building capacity and the function of the room. Even with the most efficient filter system, the deposition of fabric lints or human danders will remain an issue if there is too much human activity or circulation in the room. Table 10 shows five progressive specifications related to the performance of filter systems. When a museum has a centralized HVAC system, a final stage filter with at least 70–75% efficiency (dust spot test) should be specified. This filter performance corresponds to Level D, and represents a performance slightly above that expected for office buildings. However, at this level only about 40% of the fine particles are filtered. Subsequent classes represent more progressive, balanced performance levels. At Level B, a gas filter should be considered for more global pollutant control. Level A is the optimum level. It provides superior control of fine particles (a reduction of more than 80%, with an acceptable pressure drop). Different configurations of filters are also possible as long as the minimum performance required is respected.

Due to the high cost of high-performance filters, a lower-performance, first-stage filter is often installed upstream to extend the useful life of the high-performance final filter. The final filter must be installed downstream of the fan for optimal efficiency in positive pressure. It is also possible to add, before the first-stage filter, a prefilter for very coarse particles such as sand, insects, and dead leaves. However, the use of medium-performance first filters is more common, partly because the price of these filters has fallen but also due to the long-term benefit of postponing the need to clean the climatization unit (e.g. heat exchanger or cooling coils). Reasonable energy consumption, the extended life for the final-stage filter, and dust abatement procedures leading to finer particles with less mass have all promoted the use of medium-performance filters. Furthermore, air intakes are most often placed at the roof level rather than close to the dust-loaded ground level. The use of medium-performance filters as first-stage filters has been well documented in some engineering standards, such as the Swiss (SWKI 96-4), German (VDI 6022), and Canadian guidelines (Société Suisse 1998; Verein 1998; Nathanson et al. 2002).

With increasing concerns about viruses, bacteria, and metallic toxic dust, HEPA filters have become more popular and readily available. A true HEPA filter removes 99.97% of all 0.3-μm particles while a near-HEPA filter removes particles with an efficiency between 90 and 99.97%. Because of the high pressure drop, HEPA filters are very energy demanding. HEPA filters are mainly used for the pharmaceutical or semi-conductor industries where staff must wear special clothing and masks. Obviously, this is not

TABLE 11. GASEOUS SORPTION PERFORMANCE OF FILTERS[a]

Sorbent	Impregnant	Acetic acid	H$_2$S	NO$_2$	O$_3$	SO$_2$	Water vapour	Formal-dehyde	NH$_3$
Activated carbon	None	Good[b]	Poor	Poor	Good	Good	Poor	Poor	Poor
	Potassium carbonate (KCO$_3$) or potassium hydroxide (KOH)	Good	Good	Good	Medium	Good	None	Poor	Poor
Activated charcoal cloth	None	Good	Poor	Poor	Good	Medium	None	Medium	Poor
Activated alumina	Potassium permanganate (KMnO$_4$)	Good	Very good	Medium	Good	Good	None	Medium	Poor
	Sodium bicarbonate (NaHCO$_3$)	Good	Medium	Good	Medium	Good	None	Poor	Poor
Fixed-bed composite materials	None	Medium	Poor	Poor	Good	Good	None	Poor	Poor
(multi-layered sorbent media)	Catalysis/potassium carbonate (KCO$_3$)	Good	Good	Good	Medium	Good	None	Poor	Poor
	Metal salt	Medium	Very good	Medium	Medium	Medium	None	Medium	Good
	Ion-exchange resin	Good	Medium	Medium	Medium	Medium	None	Poor	Very good
Molecular sieves[c] (zeolite)	None	Medium	Medium	Poor	Poor	Medium	Very good	Poor	Poor
Silica gel	None	Medium	Poor	Medium	Medium	Medium	Very good	Poor	Medium
Zinc oxide catalyst	None	None	Very good				None		
Water spray (wet scrubber)		Good	Poor	Medium	Poor	Good		Medium	Medium

a: Performance in active mode. Manufacturers can provide a blend of some sorbents; their net performance will be the average performance of each single sorbent.
b: Medium performance for formic acid.
c: Highly dependent on moisture content.

Sources:
Adapted from ASHRAE (1999b); Muller (nda, ndb); Purafil (1999); Waller (1999); Charcoal Cloth Limited (1990); Kames (2000, 2002).

the case for museums offering public access to their collection. In some special cases, restricted access collections will benefit by having very high-performance filters if the use of enclosures is not an option. This performance is referred to as specification Class AA in Table 10. If the gas filter releases sorbent particles, it is recommended to include a pre-final-stage filter in front of the expensive final filter.

For controlling gas and vapour pollutants, a gas filter is added between the first- and final-stage particle filters. Table 11 shows the relative performance of some gas filters to different key pollutants based

on the main sorbent[3] medium and additional active impregnants. This comparative table can be useful for selecting the proper sorbent systems for general performance or for tight control of a specific pollutant. Unfortunately, few systematic comparisons between sorbents from various manufacturers exist. Some performance ratings are based on theoretical performance and on experts' judgment. According to the table, only three sorbents provide good general control of key gaseous pollutants (excluding water

3. Basic information on physical and chemical sorption phenomena can be found in various reference documents such as ASHRAE (1999b).

vapour). They are activated carbon impregnated with potassium carbonate or with potassium hydroxide, activated alumina impregnated with potassium permanganate, and fixed-bed filters* having a catalytically active carbon and potassium carbonate impregnant. Few filters can provide high-level performance for specific pollutants. For example, only three gas filters can capture hydrogen sulphide very well. They are activated alumina impregnated with potassium permanganate, fixed-bed filters with metal salts, and the zinc oxide catalyst. For a high-performance filter system, an HVAC engineer may suggest two-stage gas filters to improve overall performance (e.g. two fixed-bed filters: one with catalytically active carbon and one with potassium carbonate in series with metal salts). A second possibility is an activated carbon sorbent impregnated with potassium carbonate coupled with activated alumina impregnated with potassium permanganate. In some cases, special sorbents can be blended into one single gas sorbent. Besides choosing the right sorbent for the filter system, the filters should be designed to provide high single-pass efficiencies for intake air at reasonable energy costs.

The use of sorbents impregnated with strong oxidant, acid, or alkaline compounds in an HVAC system can be a concern if there is any possibility of very small amounts of sorbent powders escaping the final filter and spreading through the galleries. It has been reported that sorbent medium can corrode the metallic filter-holding trays or modules (ASHRAE 1999b). However, these filters are not new to the market and are widely used in the high technology field without noticeable problems. It seems the undesirable consequences are negligible compared to the benefit of low levels of gaseous pollutants. A close look at the nature of dust deposition and its possible adverse effects on objects is suggested. Fixed-bed filters do not have this disadvantage as shown in Figure 32.

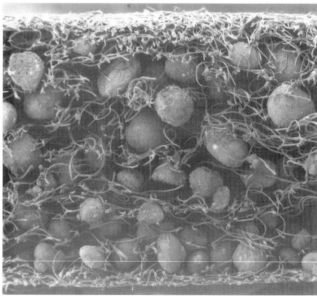

Figure 32. Fixed-bed filter. Note the distribution of sorbents within the fibrous structure and the bond between fibre and sorbent particles. Courtesy of AQF Technologies, a division of BBA Filtration.

It is usually easy to know when particle filters are saturated and need replacement: the pressure of the HVAC system drops, the filter gains substantial weight, or dust can be seen on the filter. Apart from the browning of potassium permanganate impregnated active alumina, it is much more difficult to know when gas and vapour sorbent media are saturated. Even if there is no standard protocol to determine the lifetime of these filters, there are three common methods to verify if they are exhausted: 1) metal coupons can be placed downstream of the filter for 1–3 months and then the coupons sent for

quantification of the corrosion development (Purafil 1998a, b); 2) samples of the sorbent medium can be taken and sent for analysis (the means of sampling of the sorbent is critical: if the sample is taken only at the surface, the analysis will always show a degree of saturation); 3) some key pollutants can be monitored downstream of the HVAC system. All these methods require laboratory services. The results of these analyses will indicate the service life of the filter system, which can then be changed at fixed intervals. If no testing can be done, replacing the gas filters every year is suggested.

Electrostatic air cleaners incorporated into HVAC systems are not appropriate for museums. With this type of cleaner system, particles acquire a charge as they pass through a high voltage. Negatively charged particles are then attracted by oppositely charged surfaces from which they may be removed later. However, the ionization produces substantial amounts of ozone, a strong oxidant.

A wet scrubber or water spray system incorporated into an HVAC system partly removes particles and gases by bringing them into contact with water. Such a system, by itself, does not reduce the water vapour infiltration into the building. It has an uneven efficiency for gaseous pollutants but it does reduce the level of fine particles ($PM_{2.5}$) by about 50% and particles having a diameter of more than 10 μm by more than 99% (Air Pollution Training Institute 2000). The main disadvantage of wet scrubbers is that they need periodic cleaning to avoid calcareous deposition and mould contamination.

Control with Portable Filtration Units
Portable filtration units are an efficient way to control the climate at the room level for a moderate cost. They provide, locally, good preservation of sensitive objects. The environmental performance maintained by a portable unit will be optimal if the room is well isolated, such as a collection storage room. Most particle and gas filters can be adapted for portable units.

Control with Natural Ventilation
Museums without an HVAC system can take advantage of the spatial gradients (dilution and sorption) through their corridors and rooms by placing objects vulnerable to outdoor pollutants far from entrances or windows which are periodically opened. However, this approach will not provide sufficient performance for well-visited museums where adequate fresh air is required for the health of visitors and staff.

Vacuum Cleaning
Proper housekeeping is important. When vacuuming floors be sure to use proper filter bags as vacuum cleaning can resuspend about half of the fine particles ($PM_{2.5}$) when the filter's efficiency is below 75%. This means that, in a museum without an HVAC system, each time vacuuming is done only about half the deposed fine particles will be captured while the others may settle down in hard-to-clean places such as an object's surface. Fortunately, it is now easier to get vacuum cleaners designed for high-efficiency filtration (Stavroudis 2002a, b).

REDUCE REACTIONS
Relative Humidity
Where tight climate controls are difficult to achieve, museums in countries having cold seasons should take advantage of them by maintaining a dryer environment in the room throughout the cold period (e.g. cold season RH 10% less than warm season RH) (ASHRAE 2003). This will be beneficial to most collections as shown in Figure 21. However, RH transitions must be done gradually. Before reducing the RH, objects sensitive to humidity fluctuations (such as miniatures or fragile composite objects) should be identified as they may have to be dealt with separately.

Temperature
Temperatures are set for human comfort. However, organic objects benefit when stored at cold temperatures. The main drawback is that cold temperatures necessarily imply limiting access to the object. Also, many organic objects stored below 10°C become more brittle and must be handled with care. Low-temperature rooms are mainly used for large hypersensitive collections, such as nitrate and acetate cellulose collections, colour photographic materials, and skin and fur ethnographic collections.

More information related to the risk associated with low or fluctuating RH and temperature can be found in CCI Technical Bulletin No. 23 *Guidelines for Humidity and Temperature for Canadian Archives* (Michalski 2000) or "Museums, Libraries, and Archives" (ASHRAE 2003).

REDUCE EXPOSURE TIME
Another option to minimize the rate of deterioration or risk over time is to expose objects only intermittently to pollutants. The rest of the time they should be stored in a cleaner environment.

CONTROLS AT ENCLOSURE LEVEL

Pollutants (except particles) do not usually cause rapid deterioration to objects in a room with minimal environmental control. However, this situation is not true for display cases or storage cabinets. Although enclosures have a high potential to block the infiltration of outdoor pollutants, they may also trap harmful pollutants released inside the enclosure. This can lead to a high risk of deterioration for objects as they are exposed to levels of pollutants that are higher than normally found in rooms. Pollutants within an enclosure may originate from the use of inappropriate products or from objects in heterogeneous collections that release compounds

TABLE 12. ISSUES RELATED TO AIRTIGHTNESS OF ENCLOSURES

	Leaky enclosures (≥10 air exchanges/day)	Semi-airtight enclosures (10–1 air exchanges/day)	Airtight enclosures (≤1 air exchange/day)
Pollutants generated in the room or from the outside	• Relies on the control provided in the room. • Dust deposition is a problem.	• Weak to good protection.	• Best protection against outdoor pollutants and pollutants related to human activities in the room.
Pollutants generated inside the enclosure from the objects or products	• The leaks allow levels to remain low inside.	• Levels inside will be low to medium.	• Mainly a concern for objects sensitive to carbonyls or sulphur compounds. • Can reach maximum levels (pseudo equilibrium). • Sorbents may help to reduce the levels if the rate of emission is not too high.
Relative humidity	• Will be same as the room. • Humidity in room should ideally be <60% RH.	• Risk of high localized RH if there is a cold enclosure wall. • Moisture-buffering products will have low to moderate efficiency.	• Risk of high RH or large RH fluctuations, if there is not a large surface of moisture-buffering products and if there are fluctuations of more than 5°C in temperature or a cold enclosure wall (see temperature). • Don't store damp objects in enclosures. • Don't place objects in enclosures during damp periods (≥75% RH)
Temperature	• Will be same as the room.	• Same as the room with a delay depending on the thermal insulation of the enclosure walls. • Lamps inside enclosures increase the temperature (which will affect the RH).	• Same as semi-airtight. • Keep the enclosure at least 10 cm away from an exterior (cold) wall or from the floor if the enclosure panel facing the wall or the bottom is not thermally isolated.
Insects	• No protection from risk of infestation.	• Little protection from risk of infestation.	• Usually good protection from risk of infestation (insects can perforate and go through thin plastic sheet).
Water leaks	• Weak protection (depends on the design). • Consider leakproof enclosure tops; raise enclosures 10 cm above the floor.	• Weak to good protection (depends on the design).	• Best protection.
Other agents of deterioration	• For physical forces, thieves/vandals, no major differences are observed between airtight and leaky enclosures. • For radiation, see comments about temperature. • For fire, see comments about outdoor pollution and temperature.		
Design and maintenance	• Made naturally by stack pressure (a dust screen is desired) or by a positive pressure filter system. Filtration systems are shown in Figure 39. Filters must be replaced periodically.	• Can use a positive pressure filter system. • Ensure proper air circulation between objects and any sorbents present.	• Must use sealed seams and closed cell gaskets. • Ensure proper air circulation between objects and sorbents. • Frequent need for access may limit the airtightness. • Sorbents will provide additional long-term protection.

that are harmful to other objects. The most common pollutants emitted inside enclosures from products or objects are acetic acid, fatty acids, formic acid, hydrogen sulphide, nitrogen oxides, peroxides, and sulphur dioxide. Most damage to objects inside enclosures is the result of decision-making mistakes, such as an underestimation of potential problems, a lack of information, or a lack of product control when the enclosure was built.

AIRTIGHTNESS

"Should our display cases breathe or be airtight?" The issue of making enclosures airtight to provide optimal preservation for objects displayed within them is often brought up. In the past, the answer was mainly based on either pollution or humidity factors. However, both issues should be considered along with others such as pest management. Table 12 summarizes the issues related to airtightness.

In general, there is an advantage to airtight enclosures, but two main conditions must be respected: first, no harmful pollutants should be generated inside the enclosure (pollutants can originate either from the enclosure products or from the objects within the enclosure) and second, RH fluctuations inside the enclosure must be minimal (this is related to the possible risk of elevated RH if ever there is a significant drop in temperature). To minimize the adverse consequences of RH fluctuation, moisture-buffering products (e.g. acid-free matboard, cotton cloth) should be laid out over large surfaces within the enclosure, and the daily fluctuation of temperature in the room should be less than 5°C. Enclosures should also be located at least 10 cm from an exterior (cold) wall or from the cold floor if the panel of the enclosure facing the wall or the bottom is not thermally isolated.[4] This separation distance is necessary to avoid temperature gradients inside the enclosure. If not respected, condensation may appear on the colder surface within the enclosure. When the risks related to emissive products and possible high RH cannot be solved, filtered positive pressure systems or micro-climate generators should be used. If these systems are too expensive, it is better to keep the enclosure very leaky and let the room's environment work within the enclosure.

The interior environment of display cases made with intentional 0.5- to 2-cm gaps along the edges

will be similar to the room environment, and will rely on the control facilities in place in the room. However, leaky cases will accumulate large size particles (mainly hair and fabric lint) due to the stack pressure* phenomenon. It should be recognized that the interior of a leaky display case requires more routine maintenance than that of an airtight case. It is common to observe display cases with clean outside surfaces but dust accumulation inside. The resources and handling risks related to cleaning the interior of leaky enclosures should be taken into account when this option is considered.

For travelling crates, objects should be wrapped with cotton or tissue paper and sealed in plastic bags with a minimum of free air space within the bag and the crate. These recommendations are, of course, in addition to the objects' padding and cushioning requirements. The case should be thermally isolated.

Some objects are considered "self-destructive." This is the case for acidic papers and cellulose nitrate and acetate objects, which are destroyed by hydrolysis in a process that is sped up by the accumulation of the degradation products. By keeping the enclosure dry, the decay of these objects can be reduced. The best way to maintain a dry environment is to minimize the water vapour infiltration through the enclosure and provide desiccants inside; otherwise, a positive pressure climate control system or overall control in the room should be considered to remove harmful products. The addition of pollutant sorbents is useful in preventing the accumulation of acidic products inside airtight enclosures. For even tighter control, reducing the temperature inside the enclosure should be considered.

AVOID

Avoiding sources of pollutants relies directly on the nature of objects and products in enclosures.

Selection of Products

As mentioned previously, emissions from enclosure products can have serious adverse effects on a collection. The most common products known to cause adverse effects are acidic products (such as wood products), paints, adhesives, RTV silicones, and products having sulphur-based compounds (such as some rubbers and dyed fabrics), as shown in Tables 2 and 4. The objects most vulnerable to the pollutants released by these products are lead, copper alloys, silver, and paper. To avoid the most common adverse effects, adhere to the following two basic strategies when selecting and using products.

---

4. For crowded storage areas keeping a distance of 50 cm between enclosures and walls is recommended for fire safety.

1. As much as possible, wait 3 or 4 weeks before installing objects within airtight enclosures that have been painted, sealed, or joined with liquid products (water- or solvent-based paints and adhesives) or products that are chemically cured in situ (two-part epoxies, two-part urethanes, RTV silicones) (see Figure 28). During this delay, keep the enclosure open in a well-ventilated area to assist the off gassing.

2. In general, avoid products containing sulphur-based compounds. The presence of sulphur compounds in a product can be identified with the lead acetate test as shown in Table 26, but this test will not indicate the possible concentration of sulphur compounds generated by the products in the enclosure. An alternative is to look at the material safety data sheets (MSDS). These sheets provide information related to safety and they can be obtained directly from the manufacturer or distributor. The information sheets list all major gases released during the combustion of the product. If sulphur dioxide (SO_2), hydrogen sulphide (H_2S), or sulphuric acid (H_2SO_4) are identified, the product contains sulphur-based compounds.

Incompatible Collection

Difficulties can occur when one object affects others in the same enclosure. This most often takes place when there is contact between two objects or between different components of the same object. In airborne modes, degraded cellulose acetate films can affect other films, mineral specimens containing pyrite or sulphur-based compounds can discolour papers or tarnish silver, and ethnographic objects made with oily materials can discolour photographs and paper. These types of objects should be identified (examine Tables 1 and 2 carefully) to avoid undesirable clusters of objects. Another type of incompatible object is one that has absorbed pollutants in a previous environment and then emits these pollutants in a new environment. The same scenario can happen if products or enclosures are re-used after being contaminated. This scenario is not always obvious if the previous location or use of the products is unknown.

BLOCK

The two main objectives are to block the infiltration of pollutants from outside the enclosure and to block the emissions from materials inside the enclosure.

Improving Airtightness of Enclosures

A well-made display case designed to be airtight typically has an air exchange rate of once per day.

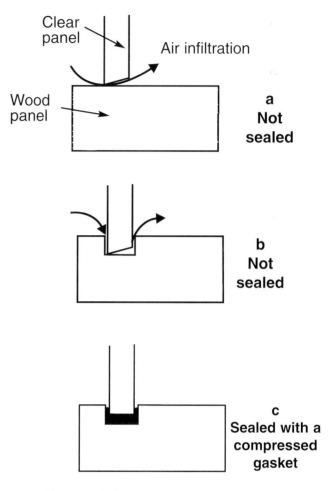

Figure 33. *Gasket compressed between two movable parts of a display case. The gasket helps to reduce air infiltration. Ideally, it should be held in a recessed channel. For a hollow gasket, the channel depth should allow for up to 40% compression of the gasket.*

This airtightness is reached by sealing the seams of the interior panel and using gaskets where there are movable parts (Figure 33). It should be impossible for a single sheet of paper to be inserted inside any crack of the case (or gaps between two parts). Gaps can be sealed, but there is a limit to airtightness due to gas diffusion through the enclosure structure. With a well-sealed display case made of acrylic windows (Plexiglas or Perspex), the major path of infiltration becomes the diffusion of pollutants through the plastic windows. Glass windows should be used instead. Ultimately, the airtightness of the enclosure will be limited by the interior pressure caused by atmospheric pressure and temperature fluctuations. Appendix 5 contains a compilation of the permeability of various films and panels to specific gases.

The Victoria and Albert Museum specifies that its display cases have an airtightness of 0.1 air exchanges per day (Cassar and Martin 1994; Martin 2000) and the U.S. National Park Service requires an air exchange rate of 0.3 for its airtight enclosures (Raphael et al. 1999). To reach such low air exchange rates, display cases made of glass instead of plastic should be considered. The benefit of this level of airtightness depends greatly on the capacity of the enclosure products to adsorb pollutants. Figure 34 shows the possible reduction in pollutants inside the enclosure, at different air exchange rates, compared with the levels outside the enclosure. These simulations are based on a deposition velocity* (or the mass transfer coefficient) of the pollutant–material systems, as shown in the following equation.

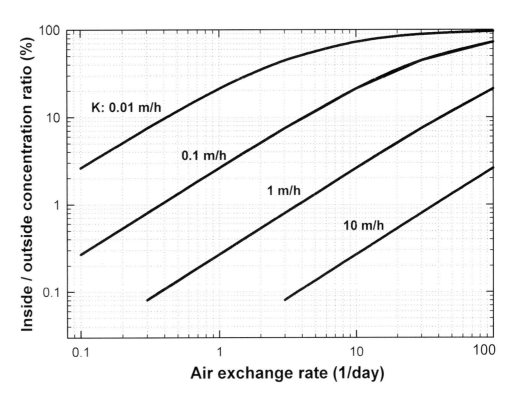

Figure 34. *Inside/outside concentration ratio (C_I/C_O) versus air exchange rate (N) of an enclosure for various deposition velocities. This simulation is based on a storage cabinet filled with objects where the objects and cabinet interior are assumed to have the same deposition velocity (K), the total exposed surface of the objects plus the cabinet (S) is 21.2 m², the net volume of the enclosure (V) is 1.37 m³, and no significant emission occurs inside the enclosure. The plotted values were obtained using Equation 1: $C_I/C_O = N/(K\,S/V + N)$.*

Equation 1 (Weschler et al. 1989)

$$C_I/C_O = N/(K\,S/V + N)$$

where C_I and C_O = inside and outside concentrations (μg m^{-3})

N = air exchange rate (1/h)

K = mass transfer coefficient (deposition velocity; V_{dep}) of a pollutant–material system (m/h)

S = surface of the material (m²)

V = net volume of air in the enclosure (m³)

Ideally, the surface and the deposition velocity of the enclosure products should be as high as possible compared to the objects. With the least sorptive products* (0.1 m/h), such as non-reactive metals or gloss-painted surfaces, a reduction of 80% in the outside pollutants is expected at an air exchange rate of once per day while a reduction of 99.7% is possible with the better sorptive products (1 m/h), such as porous materials (parameters based on Figure 34). The deposition velocities are determined experimentally and influenced by the experimental conditions. The deposition velocity tends to decrease as the product gets saturated with pollutants. This may cause an overestimation of the reduction capacity of the products at low air exchange rates. Deposition velocities found in the literature are compiled in Appendix 6.

The initial target for airtightness should be an air exchange rate of equal to or less than once per day (1/day). Until recently, measurement of the leakage rate could not be done easily. The procedure required the presence of a hole, and the determination was often done by external services. However, the author has collaborated with Micro Climate Technology to develop a simple leakage tester based on carbon dioxide measurement. This will help the makers of display cases and cabinets as well as museum staff to verify, by themselves, whether or not the airtightness of their cases fulfils the impermeability requirements

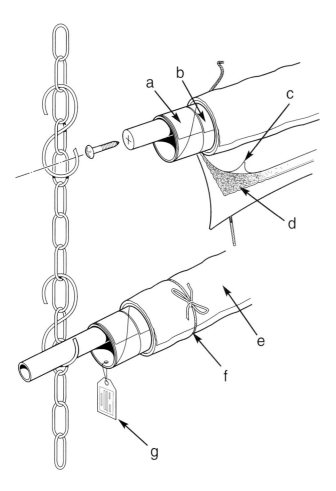

Figure 35. Textiles wrapped with cotton or acid-free tissues are well protected against airborne pollutants: (a) Mylar (Melinex) type D or type 516 covering a plastic or cardboard tube; (b) acid-free tissue or prewashed cotton sheeting over Mylar; (c) interleaving of neutral pH tissues or prewashed cotton sheeting; (d) textile with pile on outside; (e) prewashed cotton cover; (f) cotton tape; (g) identification tag.

Figure 36. A jet and space rockets wrapped in plastic during renovations.

(Keepsafe 2002a). Other non-commercial methods exist, and are shown in Appendix 7.

Protective Wrapping

When an object is in storage and does not have to be visible, wrapping it with sorbent fabrics is a low-cost, very efficient way to provide long-term protection against outdoor pollutants such as particles and oxidants. As shown in Figure 35, textiles can be rolled with two or three layers of cotton fabrics or acid-free tissues. An alkaline-buffered cardboard box is adequate to maintain low levels of outdoor acids even if the box is just moderately airtight. Silver objects wrapped with sorbent fabrics such as Pacific Silvercloth or activated charcoal cloth remain untarnished for more than 10 yrs. Even in a simple

polyethylene plastic bag, the protection of silver or any metal object is greatly improved (Figure 36). The polyethylene bag is probably the most cost-effective solution. However, the lifetime of plastic sheets for indoor use is usually limited to 10–20 yrs.

Sealing Wooden Panels

The selection and use of wood products and paints is covered in detail in CCI Technical Bulletin No. 21 *Coatings for Display and Storage in Museums* (Tétreault 1999a). Table 13 shows a simplified version of the guidelines for selecting and using coatings.

Wood product panels emit organic compounds. They may be a major source of airborne pollutants, but can be easily (although sometimes inefficiently) sealed with paints. In a ventilated enclosure, a non-emissive paint can greatly reduce the pollutant emissions from the wood substrate. However, in an airtight enclosure, as the pollutants released by the wood products diffuse slowly through the paint they

TABLE 13. GUIDELINES FOR THE SELECTION AND USE OF COATINGS

	Wood products	Metals	Concrete[a]
Enclosures such as display cases and storage cabinets	Avoid acidic woods such as oak and cedar. All paints except oxidative paints[b] are acceptable (varnishes may need many layers). All paints and varnishes must be allowed to dry for 4 weeks.	Powder coatings need 1-day drying period. With two-part epoxy or (properly) baked alkyd paints wait 4 weeks	Not commonly used.
Open structures such as storage shelves (no doors), walls, and ceilings	All paints except oxidative paints[b] are acceptable for all surfaces. Let the paint film dry for 4 days. Check with the distributor or look at the technical data to ensure the paint is appropriate for the surface to be painted. RH should be kept lower than 65% during the drying period.		
Floors (special case of open structures)	All paints except oxidative paints[b] are acceptable for all surfaces. Select paints recommended for this purpose. Let the film dry for 4 days or more if specified by the manufacturer.		
Contact between objects and paint film	For direct contact between objects and painted surfaces, wait 4 weeks after paint application. Interleaves, i.e. plastic sheets such as polyethylene and Mylar (Melinex) type D or type 516 (no polyurethane foam or PVC) or alkaline papers, can be used after 4 days of drying. For coated metal surfaces, contact between objects and paint dried for 1 day is possible with powder coatings and baked alkyd paints.		
Display and storage of lead objects in a newly painted enclosure or room	Even after selecting a suitable coating and allowing an adequate drying period, some lead objects or rich lead alloy metal objects can be altered by carboxylic acid vapours, especially acetic acid released by coatings or wood products. Avoid the use of these products for displaying or storing lead.		

a: New concrete surfaces will need to be etched by a muriatic solution to improve paint adherence. For old concrete surfaces, a TSP soap (trisodium phosphate solution, a common soap for cleaning concrete surfaces) should be sufficient.

b: Oxidative paints include the following: oil-based, oil-based urethane, alkyds, and epoxy ester (i.e. epoxy in one paint can).

Source: Tétreault (2001).

will eventually reach a significant level. It is just a question of time. It is, therefore, more effective to avoid products that release harmful pollutants than to seal them with paint. Another alternative is to apply laminated aluminum foil as a seal. This product forms an excellent gas barrier. It is known in North America mainly by the trade name Marvelseal 360, but is also available under other names. Its popularity has increased throughout the 1990s. Although not everyone likes to work with the foil, it eliminates the problem of the long drying period required for liquid paints. A low-cost alternative to commercial products such as Marvelseal can be created using aluminum foil and polyethylene grocery bags, garbage bags, or sheeting (ensure the plastic has the triangle recycling logo with the letters PE or LDPE). Assemble the barrier as shown in Figure 37, being sure to insert a paper sheet between the iron and

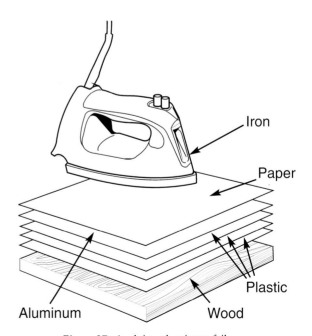

Figure 37. Applying aluminum foil and polyethylene sheets to a wood panel.

the foil to protect it from tearing. Plain aluminum foil is even more susceptible to scratches than commercial products that have a plastic film covering, so special care is important; any scratches or perforations that do occur can be repaired by adding an aluminum foil polyethylene sheet patch. Once completed, the foil should be covered with fabric or an acid-free matboard (Tétreault 1999b). Examples of the use of commercial laminated aluminum foil are described in various publications (Bosworth 2001; Phibbs 2001). It may not always be necessary to seal wood panels — it depends on the nature of the objects that will be introduced to the enclosure. Many organic objects are not harmed by emissions from wood. In some cases the use of wood may be beneficial as wood provides good control of RH fluctuations.

DILUTE, FILTER, OR SORB

When the avoid and block control measures are not feasible or not completely successful, reducing the level of pollutants inside the enclosure must be considered. Two basic approaches are possible: increasing the leakage of the enclosure (dilute) or using sorbents inside the enclosure (sorb).

The great advantage of airtight enclosures is that they reduce the infiltration of outdoor pollutants, i.e. block the pollutants. However, when the source of pollutants is mainly inside, it can be more advantageous to have a high rate of infiltration into the enclosure, i.e. dilute the pollutants. Equation 2 shows the possible reduction of a pollutant generated inside the enclosure. The equation assumes the level of the pollutant outside the enclosure is negligible.

Equation 2 (Meyer and Hermanns 1985)

$$C_{st}/C_{eq} = (K\,S/V)/(N + K\,S/V)$$

where C_{st} and C_{eq} = steady-state and equilibrium concentrations ($\mu g\ m^{-3}$)

K = mass transfer coefficient (m/h)

N = air exchange rate (1/h)

S = surface of the material (m^2)

V = net volume of air in the enclosure (m^3)

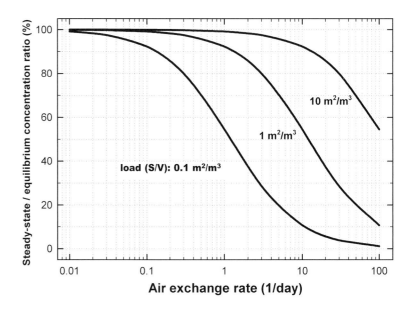

Figure 38. Reduction of a pollutant generated inside an enclosure at various air exchange rates. The plotted values were obtained using Equation 2: $C_{st}/C_{eq} = (K\,S/V)/(N + K\,S/V)$, where K (deposition velocity) was 0.5 m/h (the typical value for formaldehyde and porous products; Matthews et al. 1985), S (surface) was 10, 1, and 0.1 m^2, V (net volume of the enclosure) was 1 m^3, and there was no significant sorbent inside the enclosure.

Figure 38 is based on Equation 2, and shows the reduction in pollutant concentration at various leakage rates. An enclosure that has an air exchange rate below 0.1 per day will allow the maximum level of a pollutant to accumulate. This level is the equilibrium concentration* (some equilibrium concentrations of various pollutant–material systems are listed in Table 4). The reduction in the concentration of the pollutant that can be achieved by increasing the leakage rate of the enclosure will depend on the load (surface of the emissive product divided by the volume of the enclosure). For example, if the air exchange rate in a 1-m^3 enclosure is increased from 1/day to 10/day, the concentration of pollutants (with a mass transfer rate of 12 m/day) will be reduced by about 40% if the emissive product is 1 m^2 but will be reduced by 80% if the emissive product is only 0.1 m^2. As can be seen from this example, increasing the air exchange rate is more effective when the surface area of emissive products is small. However, although increasing the leakage rate of the enclosure can reduce the concentration of internally generated pollutants, it can also allow the infiltration of outside pollutants. Therefore, if possible, the incoming air should be filtered (e.g. with a dust screen).

A variety of techniques have been developed to filter the air from outside and to reduce the level of internally generated pollutants. Figure 39 shows three different control measures to reduce pollutants: filtration (sorption) systems that are passive, internally active, and externally active (positive pressure).

A passive system is simply a sorbent that traps pollutants by diffusion (Figure 39a); there is nothing mechanical involved. This situation can be viewed as a competition where the pollutants are reduced through selective sorption inside the enclosure; large surface areas of the sorbent compared to those of the object will minimize the sorption of pollutants by the object. Sorbents can be very effective in controlling the level of pollutants inside an enclosure, but they have traditionally been underused (with the exception of oxygen and water sorbents).

Equation 1, introduced previously, provides an estimation of the reduction of the concentration of pollutants inside an enclosure based on constant deposition velocities. However, it does not account for the fact that the sorption capacity of materials is reduced as they become saturated. To compensate for this limitation, further parameters must be considered. Equation 3, derived from Equation 4 (p. 58), factors in the sorption capacity of materials (sorption isotherm), and can be used to determine the quantity of sorbent required to maintain the pollutants inside an enclosure below a specific level.

Equation 3 $Q = (C_O V N + E S) t / (P C_I)$

where Q = quantity of sorbent recommended (g)[a]
      C_O = concentration of the pollutant outside the enclosure ($\mu g\ m^{-3}$)
      V = net volume of air in the enclosure (m^3)
      N = air exchange rate (1/day)[b]
      E = area-specific emission rate of the polluting material ($\mu g\ m^{-2}\ day^{-1}$)[c]
      S = exposed surface of the emitting material (m^2)
      t = minimum number of days the pollutant level inside the enclosure must be maintained below C_I (day)
      P = specific sorption capacity of the sorbent ($\mu g\ g^{-1}\ \mu g^{-1}\ m^3$)
      C_I = maximal acceptable concentration of the pollutant inside the enclosure ($\mu g\ m^{-3}$)

Notes:

a: Assuming the sorbent has a thickness of ≤2 cm and is well spread throughout the enclosure (otherwise a little fan may be required); the sorbent in question is the only significant sorbent material in the enclosure; and the difference between the level of the pollutant outside the enclosure and the level inside is at least 1 order of magnitude.

b: Leakage rate normalized to the pollutant of interest.

c: Ideally, this value should be determined in a situation similar to the one used in the model, i.e. similar leakage rate (N). For example, where N varies between 8 and 13/day, the following

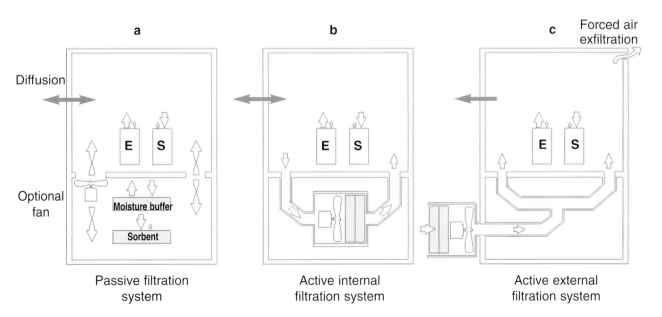

Figure 39. Design of filtration systems for enclosures. E=emissive objects; S=sorptive objects.
Note: the filtration unit in "c" can be located inside the enclosure.

area-specific emission rates have been found: a 15-year-old oak panel released 1300 μg m^{-2} d^{-1} of acetic acid; acid-type silicone cured for 9 days released 50 000 μg m^{-2} d^{-1} of acetic acid; smelly cellulose acetate film released 100 000 μg m^{-2} d^{-1} of acetic acid; and a medium-density fibreboard panel or wooden coin drawer released about 3000 μg m^{-2} d^{-1} of formic acid (Ryhl-Svendsen 2000).

In determining the quantity of sorbent required, Equation 3 considers the infiltration of pollutants outside the enclosure (terms C_O, V, and N) and the emission of pollutants from materials inside the enclosure (terms E and S). The term P is the specific sorption capacity of the intended sorbent; its unit is "μg g^{-1} μg^{-1} m^3" (μg/g per μg/m^3), which represents the number of micrograms of the pollutant that can be sorbed by one gram of the sorbent for each microgram per cubic metre of the inside pollutant concentration (C_I). The bigger the P value, the better the capacity of the sorbent to trap the pollutant. This value is specific for each pollutant–sorbent system and can be determined experimentally. Ideally, this determination should be done at (or close to) the desired range of concentration of the pollutant in the enclosure. Alternatively, the P value can be estimated from good quality sorption isotherm curves. Although P values have not currently been determined precisely for most pollutant–sorbent systems, the following values can be estimated for activated carbon in relation to some key pollutants (based on the work of Parmar and Grosjean 1989, 1991): P = 5.2 μg g^{-1} μg^{-1} m^3 for acetic acid; 7.5 for H_2S; 7.9 for NO_2; 5.2 for O_3; and 5.1 for SO_2. Activated carbon is a common and inexpensive sorbent; it has shown good performance for most key airborne pollutants and it can be regenerated for future use (unlike sorbents with active chemical agents).

The following example demonstrates the use of Equation 3:

The average concentration of NO_2 outside an enclosure (C_O) is 10 μg m^{-3}. The net air volume of the enclosure (V) is 1 m^3, it has a leakage rate (N) of 1/day, and there is no significant generation of NO_2 inside it (i.e. E = 0). The goal is to maintain the NO_2 concentration inside the enclosure (C_I) below 1 μg m^{-3} for at least a year (t = 365 days) using activated carbon (P = 7.9 μg g^{-1} μg^{-1} m^3 for NO_2) as sorbent. Using Equation 3, the amount of sorbent required (Q) to meet these requirements is:

$$Q = (C_O \, V \, N + E \, S) \, t \, / (P \, C_I)$$
$$= ((10 \times 1 \times 1 + 0) \times 365) \, / \, (7.9 \times 1)$$
$$= 460 \text{ g} \approx 500 \text{ g}$$

Therefore, 500 g of activated carbon will maintain the NO_2 concentration inside this enclosure below 1 μg m^{-3} for a period of 1 yr. After the 1-yr period, the carbon sorbent will have to be regenerated or replaced. If the C_I target was 0.1 μg m^{-3} (i.e. 10 times lower), 10 times more carbon (i.e. 5000 g) would be required.

From this example it can be seen that 500 g of activated carbon is sufficient to reduce the level of key pollutants infiltrating a moderately airtight 1-m^3 enclosure (N ≤ 1/day) by a factor of 10 compared to the outside level. This is a conservative estimate as this amount of carbon sorbent did not become saturated during the experiments of Parmar and Grosjean (1989). If the pollutant levels in a room are sufficient to meet a preservation target of 1 yr for the collection, 500 g of activated carbon sorbent will reduce the pollutants inside a 1-m^3 enclosure to a level that will meet a preservation target of 10 years for a 1-yr period, and 5000 g will lower the level to meet a preservation target of 100 years. These quantities of carbon sorbent will also control pollutants such as acetic and formic acids that are generated inside the enclosure, as long as there are no major emissive materials (E x S ≤ 1000 μg/day). To meet this condition, uncoated wood panels, fresh paints, and acid-type silicone should be avoided. Also, the internal emission (E x S) of SO_2 should remain below 1 μg/day, H_2S should remain below 0.5 μg/day, and NO_2 should remain below 5 μg/day.

To ensure optimal performance of sorbents, it is important to have good air circulation throughout the enclosure. One common design is to place the objects in the top part of a display case, and the moisture-buffering products and pollutant sorbents in the lower part. This arrangement requires sufficient openings at the interfaces to facilitate air exchange, and may include a fan to increase air circulation.

The second filtration system for reducing pollutant levels inside enclosures is an internally active air filter (Figure 39b). This type of system is based on recirculation of the air (air is blown through a series of gas and particle filters with the help of a fan). When the pollutant levels inside an enclosure are high, either from infiltration or from emissive materials, active systems are more efficient than passive systems in lowering the pollutants but they will saturate faster. The disadvantage of this type of system is the presence of electric wire in the enclosure, which increases the risk of fire.

The third system is based on a positive pressure system (or forced external air infiltration) (Figure 39c).

This type of system is used when the enclosure is not airtight or when there are important internal sources of pollutants. A ventilation system forces the air from the ambient room to go through filters before it infiltrates the enclosure (Raphael et al. 1999). The positive pressure created inside the enclosure forces the interior (enclosure or object's) pollutants to exit through leakage. It also effectively blocks the infiltration of outside pollutants through gaps. This system was found to be very efficient at avoiding dust deposition inside cases. Instead of having visible dust inside the case in 1 yr, no dust deposition is observed for more than 10 yrs. Some commercialized systems (micro-climate generators) that are designed primarily to control the RH of an enclosure work in a similar fashion (Michalski 1982; Keepsafe 2002b).

If the main source of pollutants is located inside an enclosure, a positive pressure system should be the first choice; otherwise, an internally active air filtration system would be more suitable — although in this case the sorbent can saturate quickly if the source emits a high level of pollutants. For both systems, the filtration of the lower part of display cases should ideally be excluded to ensure optimal filtration of the upper part, where the objects are located (see Figures 39b and 39c).

In addition to the sorbents typically used for HVAC systems (activated carbon and potassium permanganate impregnated alumina), there are many other sorbents capable of reducing the level of pollutants in enclosures. Porous products such as wood products, cotton, silica gel, and limestone naturally absorb specific or general pollutants. Homemade small filter systems can be made using a car dust filter in addition to Pacific Silvercloth (Figure 40) or activated charcoal cloths (Raphael et al. 1999).

The frequency with which filters must be replaced in the two active filtration systems should be based on the performance of the filter or on measurement of the saturation level of the filters as described earlier in the "Control with an HVAC System" section.

Water vapour and oxygen are sometimes the main pollutants or at least major players in the deterioration of objects, and they can be controlled by various methods. RH fluctuations can be minimized and/or actual levels reduced using either an active system (Michalski 1982; Keepsafe 2002b) or a passive method such as placing preconditioned silica gel in the enclosure. [For information related to the conditioning of silica gels at the targeted RH, consult Lafontaine

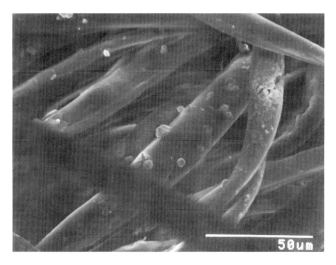

Figure 40. Microscopic detail of Pacific Silvercloth. The cloth is made of cotton fabric impregnated with fine silver particles.

(1984) or Weintraub (2002).] Oxygen levels in an enclosure can be reduced by using special sorbents or by diluting the oxygen with nitrogen or inert gases.

Controlling RH can prevent deterioration processes that occur or speed up in high humidity, e.g. some corrosion processes or migration of soluble salts into objects. [A conservator should be consulted to identify which objects require storage or display in a dry environment.] One of the most common methods to control RH in display cases is the use of silica gel. Silica gel, like organic materials, remains in equilibrium with the surrounding RH by adsorbing and desorbing water. However, there is still some confusion regarding the amount of silica gel required to control RH within an enclosure. Thomson (1977) recommends 20 kg of regular density silica gel per cubic metre, whereas Art-Sorb claims that only 0.5 kg is necessary for its products. Since regular density silica gel and Art-Sorb have similar RH buffering capacities within an RH range between 33 and 60%, the same amount of either type of silica gel should be required for comparable performance.

Silica gels can be compared by looking at their respective "specific moisture reservoir" values (Thomson 1977). This variable (M) is the amount of water (in grams) that is gained or lost by 1 kg of silica gel for each 1% change in RH. Yu et al. (2001) used this approach to evaluate the efficiency of various silica gels that are commonly used to control RH. Their results indicated that the value of M varied depending on the RH range studied, and on whether the silica gel was adsorbing or desorbing. In some cases, the difference in the adsorption and desorption curves (referred to as hysteresis) was quite significant. For example, the M value of regular density silica gel was 4.5 when adsorbing water between 33 and 60% RH, but only 3.4 on the desorption curve. Over a relatively large range of RH, it is important to account for this difference. A hysteresis corrected M value, referred to here as M_H (Weintraub 2002), can be derived using the equilibrium moisture content (EMC) on the adsorption curve at the high end of the RH range and the EMC on the desorption curve at the low end of the RH range. Using this approach and the experimental data from Yu et al., M_H values were determined for a variety of silica gels (see Table 14).

The actual amount of silica required to control the RH in an enclosure depends on more than just the specific moisture reservoir of the gel. Many variables are involved (see Equation 4). This equation was developed by Weintraub and Tétreault; it is adapted from equations in Perkins (1987) and Thomson (1977).

Equation 4
$$Q = (C_{eq} \, D \, V \, N \, t)/(M_H \, F)$$

where Q = quantity of silica gel recommended (kg)[a]
C_{eq} = concentration of water vapour at equilibrium (g m^{-3})[b]
D = decimal difference between the RH outside the enclosure and the targeted RH inside (no unit)[c]
V = net volume of air in the enclosure (m^3)
N = air exchange rate (1/day)[d]
t = minimum number of days the targeted RH range must be maintained (day)
M_H = specific moisture reservoir of silica gel, including the effect of hysteresis (g kg^{-1} %RH^{-1})[e]
F = targeted range of RH fluctuation (%)[f]

Notes:
a: For sorbents available in a sheet format, the density of silica gel (g m^{-2}) in the sheet must be known to determine how much is required.
b: The equilibrium concentration of water vapour (100% RH) will vary depending on the temperature, e.g. 17.3 g m^{-3} at 20°C; 20.0 g m^{-3} at 22.5°C; and 23.1 g m^{-3} at 25°C.
c: For example, if the average minimum RH in the air surrounding the enclosure is 30% and the targeted inside RH is 50%, the difference is 20% and D = 0.20.
d: The leakage rate is normalized to water vapour; if a precise measure of the airtightness of the enclosure is not available (Appendix 7), a value of 1 air exchange per day is commonly used to represent a typical, moderately sealed exhibit case.

e: This takes into account the effect of hysteresis within the RH range maintained within the enclosure.
f: For example, if the maximum allowable RH fluctuation is ± 5%, then F = 10.

Thomson's calculation used a formula that takes into account the fact that air exchange is more accurately described by exponential decay. However, if the greatest change in RH (from high to low) occurs in 90 days, Equation 4 gives results that are approximately the same as Thomson's. In fact, using Thomson's M value of 2 gives a result of 18 kg at 22.5°C (Thomson calculated 18.75 kg, and rounded it off to 20).

Figure 41 illustrates how Equation 4 can be used to determine the amount of silica gel required in an enclosure. The three diagonal lines represent different acceptable ranges of humidity fluctuation within the enclosure. This figure is based on the following assumptions:
- RH fluctuations inside the enclosure are based on regular silica gel with $M_H = 3$ (values in parentheses are based on double performance silica gel with $M_H = 6$)
- the RH outside the enclosure is in the range 30–70% with a maximum fluctuation of ±20%
- the RH within the enclosure is maintained at 50% within a range of F
- the thickness of the silica gel is ≤2 cm, and it should ideally be spread evenly throughout the enclosure
- the water buffering capacities of the enclosure materials (products and objects) are negligible and there is negligible spatial RH variation inside the enclosure
- temperature inside the enclosure is fairly stable

TABLE 14. SPECIFIC MOISTURE RESERVOIR

Moisture sorbents	Specific moisture reservoir, M_H (g kg^{-1} %RH^{-1})[a]
Rhapid gel	7.0
Artengel	5.7
Art-Sorb	3.7
Regular density silica gel	2.8
Wood (unpainted)	≤2
Cotton, linen, paper	≤1

a: For RH range 30–60%.

Sources: Thomson (1986); Yu et al. (2001); Weintraub (2002).

Another important factor that affects the RH buffering performance of silica gel is the nature of the objects contained within the enclosure, and their relative volume compared to the volume of air surrounding them. Organic materials are hygroscopic, and can assist in reducing RH fluctuations. Organic objects that are thick, such as a wooden sculpture, require a long time to respond fully to changes in RH. However, if there is a significant amount of organic material with a large surface, the object(s) will have a major impact on the RH within an enclosure. Any other hygroscopic materials within the case (such as cloth and wood products) will also have a buffering effect. If all the hygroscopic materials are conditioned to the targeted RH level, the buffering effect can be quite advantageous. Although silica gel has a higher buffering capacity than organic materials in the low to mid range of RH, the hygroscopic materials can serve as a supplement.

Figure 42 shows another application of Equation 4. The four diagonal lines represent different periods of time for which the specific targeted RH (20% below the average RH outside the enclosure) can be maintained within a ±5% range within the enclosure (target F of 10%). This figure is based on the same assumptions as Figure 41. If the enclosure is leaky (N > 1/day) and a low maintenance protocol is required (i.e. the silica gel will not be replaced/regenerated for at least 1 yr), the use of silica gel may not be the best option for a large enclosure. In this case, an active system may be more suitable for controlling the RH.

Objects vulnerable to photo-oxidation can be protected in an anoxic environment*. In the 1990s, an oxygen sorbent known under the trade name Ageless (Figure 43) was commonly used for the long-term preservation of rubber and some plastic objects. This oxygen sorbent is based on iron powder, and is quickly exhausted in ambient air due to the high level of oxygen. To

increase the useful life of this sorbent it is necessary to slow down the oxygen infiltration into the enclosure. This requires an adequate oxygen barrier and proper sealing (Appendix 5 provides a list of the permeability of various products to selected pollutants). When tridimensional objects are to be stored in an enclosure, it should be flushed with nitrogen or other inert gases to avoid shrinkage

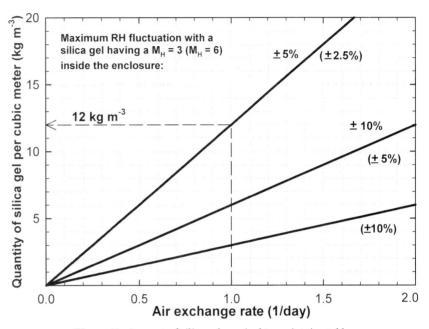

Figure 41. Amount of silica gel required to maintain stable RH conditions within a fixed range over an annual cycle.

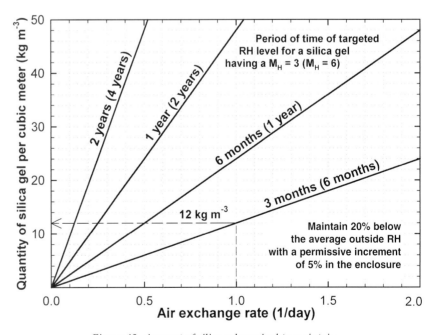

Figure 42. Amount of silica gel required to maintain a specific RH level over different periods of time.

of the enclosure volume due to oxygen reduction. Figure 44 shows an example of the use of oxygen sorbent for the preservation of a small oxidant-sensitive object. The object is enclosed in two plastic bags, which provide two good barriers. The inner bag contains the object, oxygen sorbent packets, and an oxygen dosimeter (a pink tablet that turns blue when the concentration of oxygen becomes higher than 0.5%). Between the inner and outer bags, additional packets of oxygen sorbents are added with another oxygen dosimeter. This design offers two levels of protection for the object. It is protected against possible pin holes in one of the two bags or from the slow exhaustion of the oxygen sorbents. In that case, only the oxygen dosimeter in the first level of protection will turn blue and the oxygen sorbents between the inner and outer bags can be replaced without disturbing the atmosphere of the inner bag. There are only a few examples of the use of an oxygen-free environment for objects in display cases; these include the Charters of Freedom of the United States, the Constitution of India, and some mummies in a helium or nitrogen atmosphere (Calmes 1985; Maekawa 1998). Nevertheless, anoxic environments offer great potential for preventing fading of works of art in airtight glass frames. More information on the use of oxygen sorbents and their limitations can be found in Shashova (1999). Oxygen-free environments can also be used successfully to deal with objects infested with insects (Maekawa 1998; Selwitz and Maekawa 1998; Chaumier 1998; Pacaud 1998, 1999).

REDUCE REACTIONS

The most efficient way to minimize the adverse effects of pollutants is to reduce the RH. This also reduces deterioration when water vapour is the primary pollutant. As shown in Figure 22, RH has a large impact on the deterioration of materials. Reducing the temperature also reduces the rates of deterioration from pollutants. Although valuable objects are seldom displayed in cold enclosures (Padfield et al. 1984), cold storage using commercial refrigerators is more common. The Canadian Museum of Civilization stores its

Figure 43. Ageless (oxygen sorbent) sachets containing fine iron powder covered with sea salt and a natural zeolite impregnated with sodium chlorine. The little tablet sealed in the clear plastic bag (left) is the oxygen dosimeter.

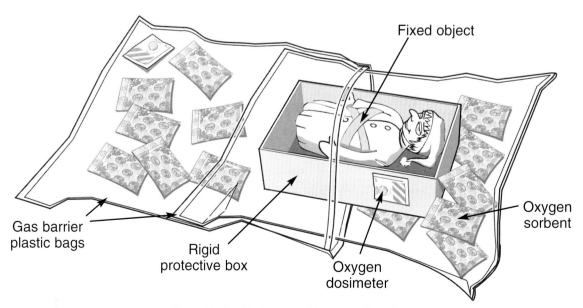

Figure 44. An object preserved in an anoxic environment.

PVC dolls in refrigerators or freezers to prevent their deterioration and the migration of the plasticizers in the PVC. Reducing either or both RH and temperature decreases the emission rate of materials. With fewer pollutants emitted in the enclosure and a lower capacity to react with objects, fewer adverse effects take place.

REDUCE EXPOSURE TIME

See the discussion on controlling building/room levels.

RESPOND

Most reported cases of sudden deterioration due to pollutants have occurred in enclosures. The problem was usually corrosion caused by emissive products. If any corrosion is noticed, all metal objects should be removed until the problem is solved. The other common problem is the discovery of a suspicious smell. An unpleasant smell does not necessarily indicate the presence of harmful compounds, but when an odour is detected it is prudent to remove vulnerable objects from the enclosure until the pollutant(s) and source(s) are identified. If the objects cannot be removed, periodic monitoring should be carried out. Resolving the off-gassing problem in enclosures is usually difficult but there are several possible strategies (Figure 45 summarizes the various mitigation solutions). The first option is to replace the emissive products or objects. If a wooden panel is the main source, it should be sealed properly. The second option is to increase the ventilation or air exchange rate of the enclosure. Also, if paints or adhesive are applied inside the enclosure they should be dried for 3–4 weeks before sensitive objects are installed. Although ventilating the enclosure is a very efficient approach to reducing the level of internally generated pollutants, unfiltered ventilation allows for the introduction of outside pollutants — which is the opposite goal of airtight enclosures. Also, it is not always easy to make display cases or storage cabinets leaky, even temporarily. The third option is the use of sorbents, and this requires that the enclosure be airtight. However, passive sorbent systems cannot work miracles. If there are high emissions of pollutants, the sorbent may not reduce

the pollutants to a satisfactory level and it may saturate quickly. The fourth option is to reduce reaction rates. This does not reduce the amount of pollutants but it does minimize their adverse effects. In some cases it is possible to combine more than one option. However, increasing ventilation cannot be combined with the use of sorbents or the reduction of RH or temperature as ventilation may introduce more moist polluted air to the enclosure than the pollutant and moisture sorbent can handle. If these options are not satisfactory, it may be necessary to reconsider the products that were used in the enclosure.

Objects or products that have been impregnated with an unpleasant smell will take a long time to desorb the vapours, especially if the object has been exposed to the vapours for a long time. Also, if the object or product is relocated to another enclosure it could lead to contamination of other objects.

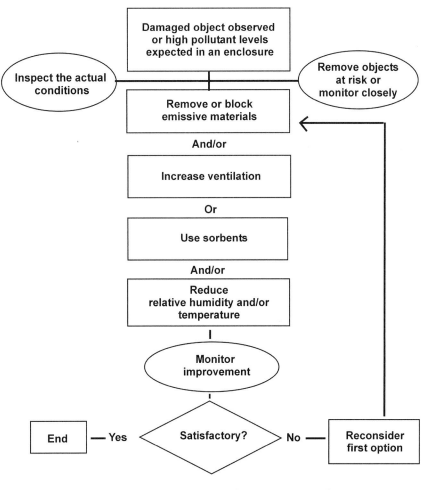

Figure 45. Mitigation options for emissive materials.

Control Strategy for Hypersensitive Objects

It is important to identify objects that are hypersensitive to pollutants (e.g. cellulose acetate- and nitrate-based objects, some colorants such as curcumin and alizarin crimson, lead, silver, natural rubbers, polyurethanes (Figure 46), objects difficult to clean, and salt-contaminated objects). Under basic environmental controls, these objects usually deteriorate at unacceptable rates. Table 15 describes typical types of damage along with some efficient control strategies and ethical considerations related to the treatment or replacement of hypersensitive objects.

Figure 46. A VHS videotape with degraded polyurethane binder. Note the flaking.

TABLE 15. CONTROL STRATEGIES FOR THE PRESERVATION OF HYPERSENSITIVE OBJECTS

Hypersensitive materials	Most harmful airborne pollutants	Control strategies[a]	Damage and ethical considerations
Cellulose acetate	Water vapour, acetic acid	• Keep dry (but not below 20% RH) (Reilly 1998) and cold. • Isolate objects that have reached the critical acidity (Reilly 1993). Provide adequate ventilation and a special vault/enclosure. Add sorbents in the enclosure (long-term efficiency is not well documented). Monitor regularly to identify those objects that have reached the fast deterioration process.	Films and sheets from the 1950s and 1960s shrink, and become brittle and sticky. If resources permit, it is best to make copies for use and store the originals at low temperature and RH.
Cellulose nitrate	Water vapour, nitrogen oxides (NO$_x$)	• Keep dry (but not below 20% RH) and cold. • Follow NFPA standards for storing large quantities of rolled films (NFPA 2001).	Films and sheets (produced mainly in the United States from 1896 to 1952) become powdery or sticky. Heavily deteriorated rolled films stored above 38°C are at high fire risk. If resources permit, it is best to make copies for use and store the originals at low temperature and RH.
Most sensitive colorants	Nitrogen dioxide, ozone, sulphur dioxide, and also oxygen + radiation. LOAED for NO$_2$, O$_3$, or SO$_2$: 1 μg m^{-3} yr	• Use airtight enclosures, store them in the dark, and display in low light intensities. • Keep dry and add sorbents in the enclosures. The most common hypersensitive colorants are alizarin crimson, basic fuchsin, curcumin, and pararosaniline base (from Appendix 2).	High probability that colorants are already partly faded if objects are usually exhibited. When half faded, the LOAED shifts up by one order of magnitude. Make a copy for frequent consultation.
Lead	Acetic acid NOAEL: 400 μg m^{-3}	• Consider enclosures that do not contain products that emit carboxylic acids or formaldehyde (wood, paint) and ensure there is no contact with papers or cardboards. • Keep dry. • Passivation of lead objects is a possibility.	Surface information can be lost quickly if efflorescence occurs. Lead objects that have a stable film layer and are clean are much more resistant to acetic acid attack.

Table 15. Control strategies for the preservation of hypersensitive objects (*continued*)

Hypersensitive materials	Most harmful airborne pollutants	Control strategies[a]	Damage and ethical considerations
Natural rubber	Ozone, oxygen LOAED of ozone for stress rubber: 0.005 μg m^{-3} yr	• Maintain low air pressure for inflated rubber objects. • Use airtight enclosures. • For long-term storage of most valuable objects, consider low-oxygen enclosures or keep cold.	Stressed rubbers are highly sensitive to ozone. Original rubbers need special care. Most original rubbers from machinery or instruments are usually replaced by new materials.
Silver	Hydrogen sulphide LOAED: 0.1 μg m^{-3} yr	• Use airtight enclosures; ensure there are no materials that emit sulphur. • For storage, wrap with sorbent fabrics. • Keep dry and use sorbents.	Tarnish removal methods often remove original silver and are a problem with thin or fragile silver surfaces.
Polyurethane magnetic tapes	Water vapour, particles LOAED of particles: 10 μg m^{-3} yr	• Use airtight enclosures and keep them dry. • Minimize handling and use of tapes. • For long-term storage of most valuable objects, keep cold.	Tapes deteriorate by photo-oxidation, hydrolysis, and abrasion of dust. If resources permit, make copies of rolled tapes for use and save the originals at a low temperature and RH.
Moisture-sensitive objects	Water vapour LOAED: 1600% RH yr for cellulose acetate	• Keep dry (but not below 20% RH) and cold (Figure 24).	Maintaining low temperature will limit access to the collection. Vibration and handling are not recommended for dry and cold objects. Some composite objects cannot be kept drier than their usual environment. Most moisture-sensitive objects stored or displayed at 50% RH and 20°C will show adverse effects after 40–50 years (15–25 years for flexible PVC).
	Moisture-sensitive objects include cellulose acetate and nitrate plastics, colour photographic prints, photographic gelatine, magnetic recording tapes, many types of papers, natural varnishes, flexible (plasticized) PVC.		
Objects that are difficult to clean	Super coarse to fine particles LOAED of PM$_{2.5}$: 10 μg m^{-3} yr	• Keep a distance of 1–2 m between objects and visitors. • Choose a filter specification of at least Class B for the HVAC system (Table 10). • Use airtight enclosures and avoid large RH fluctuations.	Particle deposition can be very hard and time-consuming to remove, and there is a high risk of damage to the object during the cleaning and handling. See below.
	Objects that are difficult to clean include those with powdery pigments or surfaces such as some painted ethnographic objects or some butterfly wings; physically fragile objects such as insect collections and filamentous mineral specimens; objects in which fine particles could become lodged in microcracks or interstices (e.g. ivories or painted objects with cracks); objects with sticky surfaces such as some deteriorated plastics and some polyethylene glycol treated wooden waterlogged objects; objects that cannot be easily cleaned by vacuum cleaning, immersion baths, or poultices; and objects with numerous small components that would be difficult and time-consuming to clean well.		
Salt-impregnated objects	Internal salt and water vapour	• In general, keep RH below 40% and minimize fluctuations; for rich soda glasses keep RH between 20 and 35%. • Avoid materials that emit acetate and formate compounds, and avoid contact with the unsealed ground. • Desalination is recommended if feasible.	Some salt or acetate compounds are inherent to the object. The complete deflagration of contaminated archaeologic glass or ceramic objects has been observed after excavation and drying.

a: Whenever possible, monitoring should be done as an integral part of the control strategy.

PRESERVATION MANAGEMENT   5

The next step in controlling airborne pollutants is to determine what can be achieved in terms of protecting a collection, and how much it will cost. This chapter proposes various ways to assess protection and to conduct a basic cost–benefit analysis.

PRESERVATION AND PERFORMANCE TARGETS

Protection can be assessed using targets based on the quality of the preservation of objects (preservation targets) or on the quality of the environment (performance targets).

- *Preservation targets*: the focus is on the object or collection and its risk or rate of deterioration.

- *Performance targets*: the focus is on the environment (assessed on the basis of the maximum concentration of pollutants, various semi-quantitative methods, or a set of specifications).

These targets are interrelated. The preservation target dictates the required performance target for the environment, which in turn dictates what control strategies can be used to meet the requirements. Unfortunately, specifications and semi-quantitative methods do not always provide predictable results in terms of environmental performance or preservation of objects. Until recently there has been little knowledge of the exposure–effect relationship of pollutant–material systems, making it very difficult to assess the degree of preservation of a collection. However, the concepts of NOAEL and LOAED allow the museum community to establish targets that offer a relevant indication of the preservation status of a collection. This information can then be communicated to the key decision-makers (e.g. senior executives and government ministries) who set the policies for protecting cultural property. The following examples are provided to better explain the application of these targets.

PRESERVATION TARGETS
The first well-known approach to establishing guidelines for pollution levels was made by Garry Thomson in the 1970s. He reported that the books in archives in rural areas of the United Kingdom were in much better condition than those in urban areas. The damage in the urban archives had occurred mainly at the beginning of the 20th century when the primary source of energy was coal combustion. Based on this observation, the average levels of SO_2 and NO_2 in the rural areas became the recommended levels for museums to ensure long-term preservation. This guideline has been used for more than 20 yrs. Guidelines based on the preservation of objects allow some freedom regarding what control strategies are used to meet the environmental targets. In the end, there is often a need to compromise between the degree of preservation desired and the means at hand to attain it. The following preservation targets start with a simple pollutant–material system and continue to a mixed collection exposed to multiple airborne pollutants.

No Adverse Effect on a Material Caused by a Pollutant
The designation of a preservation target is greatly simplified when an object is made of one single material with a well-documented NOAEL for a specific pollutant. In this case, the preservation target is zero damage, i.e. an absence of significant adverse effects by the pollutant. However, in terms of risk management, the NOAEL does not pretend to guarantee the complete absence of deterioration but, rather, provides some certainty that for an extended time there will be no significant observable or measurable adverse effect. The British Museum has done extensive monitoring of display cases made of painted wood products, and found that cases with less than 317 μg m^{-3} of acetic acid could contain lead objects for more than 11 yrs without the objects showing any signs of adverse effects (Thickett et al. 1998). If the preservation target of lead objects is zero damage, this can be met by keeping the level of acetic acid below its NOAEL. As a second example, no mould growth is observed when the RH is kept below 60%. In this case, humidity is not the direct pollutant but is the critical parameter in the development of specific airborne compounds — moulds.

Rate of Adverse Effects on a Material Caused by a Pollutant
If it becomes difficult to reduce a pollutant below its NOAEL or if the vulnerability of a material relies on the LOAED approach, its preservation target assessment relies on the empirical notion of a "reasonable" deterioration rate. This can be

expressed as one observable adverse effect per period of time. A preservation target of 10 yrs supposes there will be no sign of adverse effect for 10 yrs. The deterioration of vegetable-tanned leathers by SO_2 (Figures 47 and 48) can be used to illustrate a preservation target based on a rate of adverse effects. The LOAED of SO_2 for leather treated with tannins is 40 μg m^{-3} yr. If an adverse effect on the leather every 10 yrs is chosen as the target, the leather should be exposed to a level of SO_2 below 4 μg m^{-3} yr (assuming linear reciprocity of the dose). To respect this target, leather books located in urban museums should be exhibited in open display only 1 month per year (or 3 months every 3 yrs) and kept in airtight boxes or covered with sorbent tissues the rest of the year. It is also possible to exhibit them in glass display cases or store them in acid-free cardboard boxes. This solution should provide a preservation target beyond 10 yrs. The LOAEDs for various pollutant–material systems are provided in Table 3.

The same approach is used to protect objects against light fading. Appendix 8 shows examples for various light-sensitive colorants. The indicated light exposures satisfy the rate of observed fade per exposure period. It takes an accumulation of about 10 small observed fades (LOAED) to cause a 50% rate of fading, and after 20 observed fades the colorant should be practically all faded.

Assessing the preservation of every single object in an entire collection with NOAEL and LOAED is not

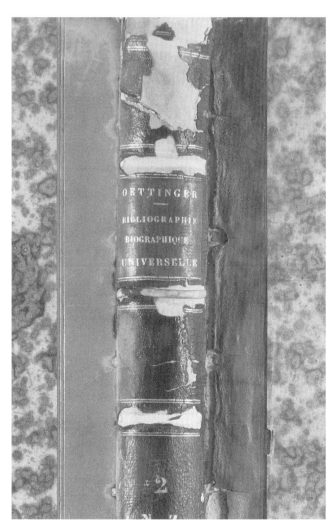

Figure 48. Another example of red rot damage. The acidic vegetable-tanned leather turned black after being in contact with water. [A colour version of Figure 48 is available on p. 95.]

Figure 47. A book covered with vegetable-tanned leather shows significant deterioration, known as red rot. This deterioration is caused by the transformation of sulphur dioxide into acid in the leather. The acidified leather surface chips off, and the underlying desiccated layer powders off when lightly rubbed. [A colour version of Figure 47 is available on p. 95.]

practical. However, such assessments can be very cost effective when applied to the most vulnerable and most valuable objects in the collection.

Rate of Adverse Effects on a Composite Object or Mixed Collection Caused by a Pollutant

This is similar to the situation described above except that there are many different materials to consider. In this case the preservation target can be chosen on the basis of the deterioration of the most sensitive material or a particular one. For example, for a watercolour drawing exposed to nitrogen dioxide, a possible preservation target could be the observation of small amounts of fading on the most sensitive colorant every 10 yrs. The determination of the preservation target for a mixed collection in the same room can follow the same logic by basing it on the deterioration rate of one key material — the most sensitive, or the most common, or any intermediate between these two. If based on the most sensitive material of the collection, it may be very expensive to meet the preservation target and some objects will be overprotected. However, if based on the most common material, more sensitive materials may deteriorate at an unacceptable rate. In this scenario, it may be advantageous to isolate the most sensitive objects and offer them special protection. The Image Permanence Institute has developed sets of useful tools for establishing the preservation performance and using a preservation target-like concept (Reilly 1993, 1998). These tools are shown in Figure 49. For predicting the fading of colour photographic films under different combinations of temperature and RH, the Institute selected 30% fade for the least stable dye as the criterion. The decision-maker has to determine, as a preservation target, the time it will take to reach this level of deterioration. A slide wheel provides some possible RH and temperature conditions as performance targets. If the time to reach 30% fading of the most sensitive dye is established as 40 yrs, the preservation target can easily be achieved at 50% RH and 21°C (see Figure 23). Any preservation target beyond 40 yrs will require tighter climate control.

A similar approach has been taken for predicting the lifetime of both new and degraded cellulose acetate films.

The application of this approach against light fading is based on a preservation target for objects with medium sensitivity (ISO 4: 10 Mlx hr). As shown in Appendix 8, less sensitive objects will be very well protected but more sensitive objects will fade 10 times faster. Some museums may consider dealing with highly sensitive objects separately.

Rate of Adverse Effects on a Composite Object or Mixed Collection in the Presence of Multiple Pollutants

There are two ways to determine the preservation targets of a mixed collection exposed to a full set of all possible harmful pollutants: micro- or macro-scale risk assessments as described in Chapter 3. The micro-scale approach refers to the individual pollutant–material system and the possible risk to the specific object in that environment (Tables 3 and 4), and the macro-scale approach refers to the vulnerability of most of the collection in the pollutant levels in the museum (Tables 5 and 6). When using a macro-scale risk assessment, the preservation target relies mainly on a "reasonable" rate of observed adverse effects

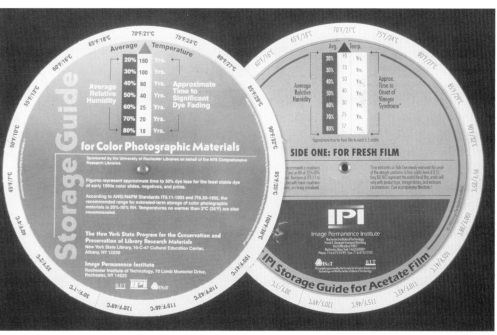

Figure 49. Life expectancy wheels. The Image Permanence Institute in Rochester, New York, has produced two circular slide rules that can be used to predict the life expectancies of colour photographic materials (left) and cellulose acetate films (right) for various temperature and RH conditions.

on the collection. The rate of deterioration chosen will dictate the average maximum levels of each key pollutant. When relying on a macro-scale risk assessment, it is important to identify all hypersensitive objects (Table 15) and provide special control measures for them as their protection will not be assured by the general classes of preservation targets.

Preservation targets can also be established as a function of the roles of the various rooms and enclosures, as shown in Table 8. Preservation targets do not have to be uniform for all objects in the collection. Instead, collections can be categorized based on value criteria developed by the museum. In that way the very valuable objects can receive closer control. Monitoring is recommended to ensure that the level of key pollutants (actual performance) matches the preservation target. As already mentioned, it becomes meaningless to seek extremely low concentrations of a pollutant for long exposure periods if that pollutant's effects become negligible at these concentrations in comparison to other deterioration processes, such as photo-oxidation, hydrolysis, and thermal degradation, which often become the dominant factors for organic materials. This would add extra cost for little or no benefit.

For agents of deterioration (such as fire, flood, or thieves) whose actions are sporadic, the preservation target of the collection can be based on the frequency of events (adverse effects). The projected frequency should be based on data provided by experts on the respective agents. For example, if a collection in a specific location statistically loses 0.4% of its material (or value) every 100 yrs due to fire, and a significant adverse effect is defined as a loss of 0.1%, the preservation target of this collection will be 25 yrs. That does not necessarily mean there will be a small fire every 25 yrs, but that is a probability. If the museum improves the building by adding fire-resistant walls or a sprinkler system, the probability of fire will be reduced and the preservation target will be improved.

PERFORMANCE TARGETS

The increasing popularity of centralized HVAC systems in the 1980s resulted in a new way to define the permissible level of some outdoor pollutants in museums based on the possible performance of the HVAC system. "Use the best technology available" or "the lowest levels possible" were both proposed, and maximum levels were assigned to some traditional museum pollutants based on typical HVAC filter performance and on the limit of detection of the monitoring technology. However, this approach has limited meaning without performing a cost–benefit analysis as the required performance targets can vary at different gallery locations due to the nature, value, or use of the collection, or simply due to the filtration capacity of the building. Within any given location, the decision-maker may use enclosures to improve the performance targets.

Performance targets can be divided in three groups: targets based on the maximum average concentration of pollutants in the ambient air surrounding collections obtained by quantitative methods; targets obtained by semi-quantitative monitoring methods; and targets based on specifications.

Performance Targets Based on the Maximum Level of Pollutants

When the NOAEL is known for a pollutant–material system, this concentration can become the performance target required for its long-term preservation. For example, a level of acetic acid below 317 μg m^{-3} is The British Museum's performance target for lead objects in display cases (p. 65). This performance target provides long-term preservation for the lead objects.

The required air quality targets for various pollutants to meet different preservation targets are given in Table 5. This table can serve as a reference for the maximum allowable level of pollutants when there is no specific knowledge available for a particular collection. Another alternative is to target the maximum allowable pollutant levels empirically.

The information in this book is not intended to set standards for museums in regard to environmental performance targets. Such standards must be decided by legal or government authorities, and their large-scale implementation would be very difficult given the different resources and constraints of individual museums.

Performance Targets Based on Semi-quantitative Monitoring Methods

Semi-quantitative methods offer an alternative to expensive quantitative monitoring campaigns. Semi-quantitative methods rely on the quantification of one or many pollutants based on an empirical scale. Some of these methods are shown in Chapter 6.

Methods based on metal coupons are popular for evaluating the air quality in enclosures. Lead, copper, and silver coupons are typically used, and alterations due to pollutants in the ambient air are monitored by visual observation, weight gain, or thickness as measured by electrochemical processes. These metal sensors are very efficient in determining whether or not the environment is corrosive. However, without specific analysis, they give little indication as to which pollutants are present and at what concentration — although copper and silver coupons are very good sensors for hydrogen sulphide due to their small LOAED value.

ISA standard S71.04-1985 "Environmental Conditions for Process Measurement and Control Systems: Airborne Contaminants"(ISA 1986), which is used to verify the efficiency of an HVAC system's filtration unit, can be helpful in assessing the results of semi-quantitative monitoring methods as shown in Table 16. For example, the cleanest category for copper corrosion in this standard is Class G1. This class (which corresponds to maximum levels of 4 μg m^{-3} for H$_2$S, 30 μg m^{-3} for SO$_2$, 100 μg m^{-3} for NO$_2$, 4 μg m^{-3} for O$_3$, and 50% for RH) can be roughly related to a preservation target of 1 yr in Table 5. Even when conditions meet Class G1, dust deposition may remain an issue. Although there is no direct correlation between these ISA classes and preservation targets, the corrosion of metal coupons can be a useful method for providing an indication of improvement after a change in control strategy or for semi-quantifying different environments.

TABLE 16. ISA ENVIRONMENTAL CLASSES

Copper coupon film thickness (nm)	ISA classes			
	G1	G2	G3	GX
	0–29	30–99	100–199	≥200
Gas concentrations (μg m^{-3}), RH <50%				
H$_2$S	<4	<10	<70	≥70
SO$_2$, SO$_3$	<30	<300	<800	≥800
Cl$_2$	<3	<6	<30	≥30
NO$_X$	<100	<200	<2000	≥2000
NH$_3$	<400	<7000	<20 000	≥20 000
O$_3$	<4	<50	<200	≥200

Source: ISA (1986).

TABLE 17. PURAFIL ENVIRONMENTAL CLASSIFICATION

Silver corrosion		Air quality classification	Copper corrosion	
Amount of corrosion (nm/30 days)	Class		Class	Amount of corrosion (nm/30 days)
<4	S1	Extremely pure	C1	<9
<10	S2	Pure	C2	<15
<20	S3	Clean	C3	<25
<30	S4	Sightly contaminated	C4	<35
≥30	S5	Polluted	C5	≥35

Source: Purafil (1998a).

Corrosion thickness can be measured with a sensor known as OnGuard which uses a copper- or silver-coated quartz crystal (increasing corrosion thickness is monitored by the decreasing frequency of the coated quartz) (see Table 24). This system can provide real-time and network readings, and the rates of tarnish film thickness.

The corrosion film thickness can also be determined by cathodic/electrolytic reduction (Purafil 1998a). After the samples are exposed, they must be sent for analysis. The results are reported on the scale shown in Table 17. Class C4 is designated as "slightly contaminated," and corresponds roughly to a preservation target of 1 yr in Table 5. Class C1, the "extremely pure" category, is equivalent to a preservation target of about 4 yrs for most collections.

Performance Targets Based on Specifications
A performance target based on specifications relies on an accurate description of the technical requirements for the performance of the building features, portable fittings, or procedures. This approach is popular because it is easier to meet specifications for individual machines, products, or procedures than to ensure performance or preservation targets. However, it is difficult to correlate a specification with a performance target. The quality of air pushed from the HVAC system through the air diffusers may not necessarily be what is monitored in the middle of the room. For example, a high specification for the performance of the HVAC system, such as Class A from Table 10, may not necessarily sustain low levels of pollutants in a room if the building is leaky and often crowded with visitors. Fortunately, at the enclosure level, specifications such as "The product must be exempted of sulphur compounds according to the lead acetate test" (Table 26) or "No vulcanized rubbers, urea-formaldehyde based glue wooden panels and paints formed by oxidative polymerization should be used" are very reliable. At the HVAC system level, it is important to prohibit the use of corrosion inhibitors based on amine compounds, as these will result in a white film deposit on the surface of objects.

Targets based on specifications can help to avoid common undesirable scenarios such as the presence of highly emissive products or high amounts of unfiltered air entering the building.

COMBINATION OF TARGETS
In practice, different preservation and performance targets can be used at the same time. Specification

targets combined with performance targets based on the measured pollutant levels or on semi-quantitative methods can be as good as a single preservation target, since it can be difficult to verify the actual value of the target (in years).

Cost–Benefit Analysis

Before any improvement in the preservation of collections is made, different options should be considered. Each one will have advantages, limitations, and short- or possibly long-term impacts on the budget. A final decision can then be made after considering the various scenarios or options, some of which may resolve only some aspects of the problem. It will often be necessary to compromise between what the decision-maker really wants, what the ideal solution is, and what can actually be done with a minimum of uncertainty.

A cost–benefit analysis is an analytical method designed to evaluate different scenarios in terms of their likely cost and potential benefit. Such an analysis can provide a systematic approach to the decision-making process. There are several well-known methods for conducting a cost–benefit analysis, some of which have been established since the 1950s. One example is the classic method proposed by Charles Kepner and Benjamin Tregoe (1976) in *The Rational Manager* (originally published in 1956). This method is the inspiration for the approach used in the simple case study discussed below. Alternative methods such as the decision-tree (Ashley-Smith 1999; Caple 2000) also exist.

Before starting an extensive cost–benefit analysis of the control of pollutants, it is important to determine if pollution control is a high priority compared to the control of other agents of deterioration, e.g. prevention of fire or improvement of security. For assistance in evaluating the risk of the different agents of deterioration, principles and guidelines that have been adapted for the conservation field should be consulted (Ashley-Smith 1999; Michalski 1994a; Waller 1999).

There are seven steps in a cost–benefit analysis:
1. Define the problem or context.
2. Set and weight the criteria.
3. Develop options and evaluate them against the weighted criteria.
4. Compare the costs and benefits of the options, and make a tentative decision.
5. Assess the adverse consequences of the tentative option.
6. Control the adverse consequences of the final decision.
7. Monitor and re-evaluate.

Define the Problem or Context
The first stage is to define the problem or context; to solve any problem efficiently, it must first be specified or described precisely. Questions such as what, where, when, or how can be used. In our case study, the problem and context can be defined as follows: an historical house where the concern is dust deposition inside four old-fashioned, leaky display cases in which small fragile objects that are difficult to clean are permanently displayed.

Set and Weight the Criteria
Defining and weighting criteria is a very important step as the criteria are the reference points on which the decision-making process will be based. Criteria usually derive from concepts associated with results and resources. They should reflect, as much as possible, the institutional objectives in terms of time, place, and quantification of a desired performance. In the field of cultural property, there are a number of well-recognized principles (some of them are described in Box 6).

A variety of professionals (e.g. curators, educators, conservators, and managers) should be involved in generating the criteria. Someone from a government heritage or preservation policy agency could be invited to provide an additional perspective. Start by holding a brainstorming session to generate possible criteria. After the ice is broken there will probably be a lot of good ideas. Group the suggestions to eliminate redundancy. When finished, there should be no more than about 10 criteria.

Once the criteria are identified, it is necessary to assess their relative importance. When the issue is simple and there are only a few participants, this can be done through an informal exchange on the various options leading to a consensus. When the issue is complex and could have a significant impact on the budget or on the institutional goal, the analytical hierarchy process (AHP) is recommended. AHP is a mathematical model that includes and measures all tangible and intangible, quantitatively measurable and qualitative criteria. Developed in the early 1970s by Thomas Saaty (1990), the model calibrates criteria into a numerical scale with a sequence of compared pairs of criteria. A simple AHP program can be found on the

CCI Web site (CCI 2003). Participants are asked to rank, on a scale from 1 to 9, how one criterion is important compared with another one. The mathematical model gives a relative weight for each criterion. The summation is normalized to 100%. The model also provides a consistency ratio, which shows how consistent the evaluation of the criteria was during the comparison. A low value is proof of good consistency. Before starting the exercise, it is important to ensure that all participants understand the meaning of each proposed criterion.

Table 18 shows examples of some possible weighted criteria. The criterion related to the health and safety of the staff and visitors has been included in the list as a must. If an option doesn't fulfil this mandatory condition, it is rejected. In our case study, this mandatory criterion is easily respected. The same criteria can be used for many different cost–benefit analyses as long as they fulfil the purpose — although they may need to be weighted differently for different purposes, e.g. aesthetics would be rated higher for exhibition than for storage.

Box 6.
Principles of preservation

For any specific problem, different people will have different answers depending on their experience, values, and principles. However, there are some basic principles that many heritage communities agree to consider closely for all cultural property.

Integrity, Access, and Preservation
Figure 50 illustrates the three main (and interrelated) principles associated with cultural property:
- the integrity of the object should be respected as much as possible
- the object should be presented (access) as much as possible
- the object should be preserved as much as possible (access for future generations)
In many cases, it is impossible to fulfil all three conditions at the same time. Depending on the scenario and the people, these three principles will be ranked differently. No universality of values is expected. Some societies will put emphasis on one criterion; others will tend to be more diversified.

Prohibition of the Preservation of One Object to the Detriment of Others
There are situations where focussing on the preservation of one object puts others at increased risk. One example of this is the preservation and display of objects in historical houses. Keeping the RH of an historical house at 50% during the winter can be very good for the furniture but can progressively damage the building's structure. Article 5 of the *New Orleans Charter* adopted by the Board of Directors of the American Institute for Conservation of Historic and Artistic Works (AIC) and the Association for Preservation Technology International (APT) covered this issue by stating: "Measures which promote the preservation of either the historic structure

or the artifacts, at the expense of the other, should not be considered."

Precautionary Principle
When there are uncertainties about the adverse consequences of a control strategy, prudence suggests the risk of a possible short- or long-term adverse effect should not be overlooked. When the sensitivity of objects to airborne pollutants is not well quantified, it is prudent to consider these objects to be more sensitive than average. But not all objects with an unknown sensitivity should be classified as hypersensitive objects. A conservator should be consulted to minimize the uncertainty. Precaution is also required when initiating a control strategy. A new method doesn't have to be banned until it can be proven harmless, but the possible impact on the collections should be monitored more closely than would be the case for a tried-and-true strategy. Some mitigation strategies should be available in the event that adverse consequences are observed.

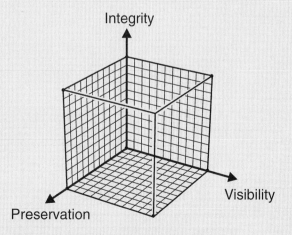

Figure 50. Principles of preservation.

Table 18. Decision analysis worksheet

Context: improve the protection of small objects in four display cases against outdoor pollutants. Criteria (as an example)	A Weighted criteria (%) (CR: 0.09)[a]	Option 1 Leaky display cases (status quo)		Option 2 Portable air filters with leaky cases		Option 3 Retrofitting the cases (make them airtight)		Option 4 Make new airtight cases	
		B Score (1 to 5)[b]	C = A x B Weighted score	B Score	C Weighted score	B Score	C Weighted score	B Score	C Weighted score
Maintain the value of the object (preservation target of 10 yrs or more)	20	2	40	4	80	5	100	5	100
Keep accessible to public (visibility, education, etc.)	20	3	60	3	60	3	60	3	60
Respect integrity of the objects	20	3	60	3	60	3	60	3	60
Respect integrity of the historical building	20	3	60	3	60	3	60	3	60
Long-term solution (>10 yrs)	7	2	14	4	28	5	35	5	35
Low maintenance (>1 yr)	6	2	12	3	18	5	30	5	30
Easy access for cleaning, upgrading, or monitoring	4	2	8	2	8	2	8	2	8
Model for the museum community (innovation, good practice, etc.)	3	2	6	4	12	4	12	4	12
Health and safety of the staff and visitors	A must	yes		yes		yes		yes	
Total of the scores (maximum of 500)			260		326		365		365

a: Consistency ratio obtained from the analytical hierarchy process. Ideally, it should be below 0.10.
b: Performance score: 1 = very poor; 2 = poor; 3 = neutral; 4 = good; and 5 = very good.

Table 19. Cost–benefit analysis worksheet

Context: improve the protection of small objects in four display cases against outdoor pollutants. Options:	Estimated cost (CAN$)	Priority		
		Weighted score		Ranking order
		(max. 500)	(%)	
1. Status quo: Four leaky display cases in the gallery	0	260	52	3
2. Keep the leaky cases and add a portable filter unit in the room (need to change the filter twice a year, electricity cost)	2000 + 150/yr	326	65	2
3. Retrofit the four leaky cases to make them airtight	1000	365	73	1
4. Make four new airtight display cases	5000	365	73	1

Develop Options and Evaluate Them Against the Weighted Criteria

When facing a specific problem, a variety of options (preventive or contingency actions from the control strategy) should be proposed and evaluated. This exercise does not necessarily involve the same people as those involved in choosing and weighting the criteria.

Begin with a brainstorming session to generate options and reduce them to a reasonable number. Once the options have been chosen, each one must be evaluated against the weighted criteria. This should be done one option at a time, as follows:

- Each participant judges the performance of the option against the first criterion using scores of 1 to 5 (1 is very poor performance and 5 is very good). The group score for the option vs. first criterion can be obtained by consensus or by averaging the individual scores. In some cases, it may be necessary to consult an expert on a specific assessment.
- The group score of the option vs. first criterion is then multiplied by the weight of the first criterion to obtain a weighted score.
- The option is subsequently evaluated against all the criteria in the same manner so that a weighted score is obtained for each criterion.
- The weighted scores for each criterion are then added together to obtain a final score for the option.

In this way a numerical score is obtained for all the proposed options. This procedure is demonstrated in our case study.

Four options are identified for the leaky display cases: keep the status quo (keep the leaky display cases); add a portable filter unit in the room; retrofit the four leaky cases; and build new display cases. The scores for these different options are shown in Table 18. These scores are repeated in Table 19, where they are expressed as percentages and ranked.

Compare the Costs and Benefits of the Options, and Make a Tentative Decision

After evaluating the options against the criteria, the option having the highest score is presumably the best solution. In some cases it may simply be the "least worst" scenario. At this stage, this option is a tentative decision. Now its feasability must be carefully reconsidered in light of its direct and indirect costs (maintenance, electricity, etc.). With our example (Table 19), options 3 and 4 both have the same score, but the retrofitting option is cheaper than buying new display cases. When one of the best options is also the cheapest solution, the choice

becomes quite simple. However, when the best solution is the most expensive, decision-makers must judge if finding the necessary funding is really the best approach. It may be that a second- or third-best option could be a suitable alternative (it must be remembered that the final score is dependent on what criteria were chosen and how they were weighted; with different criteria or different weighting, the best option might have turned out to be quite different). There is another important element of this exercise that participants must keep in mind: even after an exhaustive analysis, someone at higher levels of management may decide to choose a completely new solution. To avoid this frustrating scenario, any potential decision-makers should be included in the cost–benefit analysis process.

Assess the Adverse Consequences of the Tentative Option

Before proceeding with the tentative decision it is important to evaluate any potential risks, e.g. problems related to people, the organization, external influences, facilities, or money. These problems should be assessed based on their seriousness and probability, as shown in Table 20. For each potential adverse consequence, possible mitigation strategies should be explored. With our example, the potential adverse consequences include the risk that the display cases cannot be made completely airtight and the possibility of elevated levels of indoor pollutants within the display cases. Mitigating these risks would require monitoring the airtightness of the display cases, identifying possible emissive products, and possibly adding some sorbents.

Control the Adverse Consequences of the Final Decision

Adequate mitigation strategies must be put in place to prevent or minimize the impact of any adverse consequences of the final decision.

Monitor and Re-evaluate

Regular monitoring of the actual benefit of the chosen option should be conducted as there is often a difference between what is expected and what is obtained. If the adopted solution is not satisfactory, the problem or context should be re-evaluated.

Assessment of the Uncertainties

The example used here to illustrate the cost–benefit analysis was quite simple and provided a clearly feasible option. However, in reality, decision-making is often much less obvious. Even when the options are evaluated with a high level of competence and good judgment, some level of uncertainty remains.

TABLE 20. POSSIBLE ADVERSE CONSEQUENCES WORKSHEET

Defined risks related to Option #3 (retrofitting the leaky cases)	Seriousness (1 to 5)[a] A	Probability (1 to 5)[a] B	Priority A x B	Mitigation
Bad retrofitting; not airtight enough	5	3	15	Measure leakage rate; it should be below 1 air exchange per day and no visible cracks should be observed.
Lack of time or resources to retrofit cases	4	1	4	Plan a budget and dedicate staff time.
Condensation in the cases	2	1	2	Include humidity sorbent in the cases.
High levels of pollutants generated inside the cases	3	3	9	Assess the risk of adverse effects of off-gassing from products present inside the cases. If necessary, keep the cases drier or use pollutant sorbent. Remove objects at high risk.
Disruption of the exhibition	1	5	5	Retrofit cases one at a time. When the museum is closed for the day, empty a display case and bring it to the workshop for retrofitting.

a: Scores: 1 = very low; 2 = low; 3 = medium; 4 = high; and 5 = very high.

Any experimental measurement or judgment has some degree of uncertainty. In the example of the four leaky display cases, if the error or the uncertainty for each performance score (column B in Table 18) was estimated at ±0.5, the total weighted score for the status quo

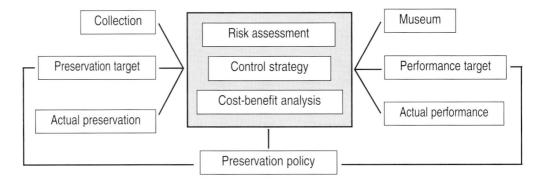

Figure 51. Collection preservation management.

(260 ± 50) would not be significantly different from the total weighted score for the retrofit option (365 ± 50). Assuming the status quo was unacceptable and some improvement had to be made, new options or weighted criteria would have to be proposed or the evaluation (or uncertainties) of the chosen options vs. each criterion would have to be reconsidered.

COLLECTION PRESERVATION MANAGEMENT

Figure 51 synthesizes the information presented in this book. It illustrates the interconnections between risk assessment, control strategies, and cost–benefit analysis, and their impact on degree of protection of the collection. It can be used to analyse the impact of change to one element on the other elements.

The analysis can start with a goal, such as the preservation of objects or the performance of the environment, or a specification (control strategy), and lead to a consideration of the cost–benefit impact of any defined element or target decision. These preservation or performance targets and their rationale should become part of the institutional preservation policy. Examples of issues that can be analysed are provided below.

- For the given preservation target (e.g. an adverse effect to a metal collection every 10 yrs), what is the maximum allowable level of pollutants? What control strategies could be used to achieve the performance target? Are they feasible within the building facilities? Are the necessary resources available? Can the required preservation target be reached?

Box 7.
Priority preservation action index

To determine which objects need primary actions, a logarithmic scale based on the empirical value of an object (or a collection) multiplied by its rate of deterioration is proposed. The value of the object or group of objects is ranked on a scale from 1 to 100 (with 100 being the highest value). The rate of deterioration is based on the number of adverse effects caused by an agent of deterioration in 100 yrs in a specific environment or context. Intermediate values between decades are also accepted for both value and deterioration scales. An object having a very low rate of deterioration (or NOAEL) should get a score of 1. In very few cases, the rating of value or the rate of deterioration can be either below 1 or higher than 100 but never zero.

The following examples demonstrate how to use the priority index.

Example 1.

An object has been ranked as one of the most valuable objects in the museum and its value score is 100. The object's rate of deterioration caused by a pollutant has been estimated as 1 adverse effect every 10 yrs so its rate of deterioration score is 10 (10 adverse effects over a period of 100 yrs). The product is therefore 1000 (100 x 10) and the index value is 3 (the logarithm of 1000 is 3). When the index is applied to different objects or collections in the museum, a trend should be observed on which objects should receive priority action to improve their preservation. Objects having an index of 3 (score of 1000) or higher need attention while objects having an index of 1 (score of 10) or below should not be considered a priority.

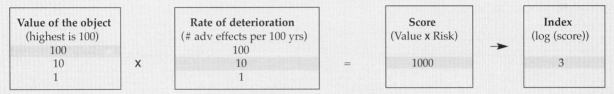

Value of the object (highest is 100)		Rate of deterioration (# adv effects per 100 yrs)		Score (Value x Risk)	Index (log (score))
100		100			
10	X	10	=	1000	3
1		1			

Example 1. An object exposed to one pollutant in a specific context.

Example 2.

This example shows how the priority index can be used to consider more than one agent of deterioration. In this case, the rate or risk of deterioration of the object must be evaluated separately for each agent. Because defining adverse effects (or loss of value) for each agent and assessing the rate deterioration in a specific environment are not always obvious, it may be necessary to consult experts. In this fictitious example, the index for each agent shows that, in this specific context, the object needs improved protection against water damage before protection against pollutants. The total index (log(total score)) can be used to compare this object's need for preservation with the needs of other objects.

Value of the object		Agents of deterioration	Rate of deterioration		Score	Index
		physical forces	10		300	2.5
		thieves, vandals, displacers	3		90	2
		fire	3		90	2
30	X	water (liquid)	30	=	900	3
		pests	10		300	2.5
		pollutants	10		300	2.5
		radiation	10		300	2.5
		incorrect temperature	1		30	1.5
		incorrect RH	10		300	2.5
					Total score	**Total index**
					2610	3.4

Example 2. An object exposed to various agents of deterioration in a specific context.

- For a given level of some key pollutants, what is the possible preservation target for a mixed collection? Which objects would be at higher risk of deterioration?

- How much care should be given to the most valuable objects compared to the overall collection? What should their respective preservation targets be, and how could these targets be achieved?

- How much will the preservation target of the collection be improved if one aspect of the building or fitting is upgraded? Is the benefit substantial enough to warrant the cost of the investment?

While many analyses can be done, it is important to focus on improving the preservation of the collection and identifying the highest-priority objects. The priority preservation action index can be used to assist in determining the priorities. This index is based on the value of objects and their risk of deterioration. More details about the index are given in Box 7. During a cost–benefit analysis, giving criteria related to the highest-priority objects a heavy weighting will allow for the most significant improvement.

NOAEL and LOAED are new tools in conservation for quantifying an object's risk of adverse effects due to airborne pollutants or other agents of deterioration. These tools are useful for preservation or performance assessments, and can provide substantial proof of the need for improvements. Cost–benefit analyses can assist in selecting the most appropriate control strategy to reach a new preservation or performance target. Collections of objects of similar value should have similar degrees of preservation applied to all agents of deterioration.

Monitoring is an important element of the control strategy. It can provide information about the environmental performance, the degree of preservation of collections, or the cause of any damage. Methods for monitoring the air quality in museums have not yet been standardized, and are not extensively used. In the past, the lack of sensitivity and the high cost of some techniques have discouraged monitoring on a large scale. There was also no clear correlation between the level of pollutants and their impact on the collection. This situation should improve in the future. This chapter provides guidelines for general and specific surveys, with an emphasis on an investigative rather than a test approach.

INVESTIGATIVE APPROACH

There are many pollutants, and there are many ways to qualify and quantify them. No single method or test can provide a complete picture of the situation. As running tests can be quite expensive and time-consuming, it is important to conduct an investigation* of the nature of the damage and the potential sources of pollutants before monitoring begins; this will narrow down the number of elements to be monitored as much as possible. Monitoring should focus on the most probable or most critical (key) pollutants. An indoor air consultant or a conservation scientist should be

contacted to select the pollutants to be monitored and the methods. Although it can provide useful data, monitoring is not always mandatory.

New monitoring techniques will soon replace those developed in the 1980s and 1990s. However, regardless of the methods used, the protocol for the investigation and the format of the report should be standardized. As a guideline, Table 21 provides information on the different stages of investigation. This protocol can be useful in comparing the level of pollutants from one site with those on other sites or with future investigations.

MONITORING TECHNIQUES

Table 22 shows various monitoring techniques grouped by their typical sampling time — ranging from a few seconds (for direct readings) up to a few weeks. The table shows the expected results and limitations of each technique for different sampling periods. More details on the various monitoring techniques, along with references, are provided in Tables 23 and 24. These tables show techniques that quantify levels of specific pollutants and those providing a semi-quantitative evaluation of one or a group of pollutants. Some of these quantitative and semi-quantitative methods are illustrated in Figures 52 and 53. Some semi-quantitative techniques are well designed for specific purposes, and are quite

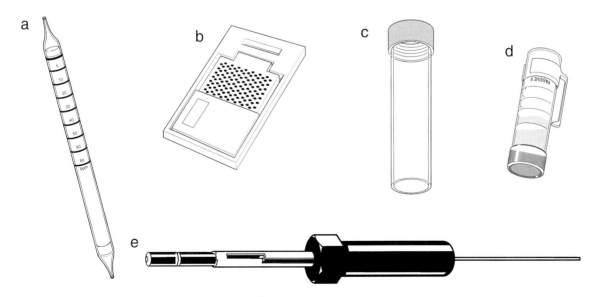

Figure 52. Quantitative monitoring: (a) Dräger tube; (b) diffusion sampler; (c) diffusion tubular sampler; (d) liquid reagent sampler; (e) SPME syringe.

accessible (e.g. the A-D strip for the determination of the critical free acid content in cellulose acetate films).

Before using expensive monitoring techniques, low-cost methods using a short sampling period (a few minutes to a few days) should be considered for preliminary assessment of the presence of abnormal levels of indoor-generated pollutants. The final choice of monitoring technique often will be determined by parameters such as the expected level of pollutants, the number of sites to be surveyed, and the time available. Tables 21–24 should be carefully examined before making a decision.

Monitoring inside small enclosures is not always easy due to the limited volume for air sampling. Opening the enclosure disturbs the concentration of pollutants, which must be allowed to return to their steady-state levels before carrying out the short-term sampling. Depending on the materials present and the volume size of the enclosure,

it may take a few hours to a few days to reach 90% of the steady-state concentration. Ignoring this fact will cause an underestimation of the actual level of pollutants.

GENERAL INVESTIGATION

A general investigation can play a role in prevention by helping to establish the overall performance of the building, the enclosures, and the procedures. As a preliminary step, the information in Tables 1, 4, and 6 can provide a rough idea of the levels of pollutants in the outdoor environment, in a room, or in enclosures, and can help in establishing possible preservation targets. The level of outdoor pollutants can also be obtained from public environmental agencies. Many of these agencies provide daily levels and trends of pollutants for major localities, and these data are often available on the Web. However, different agencies use different air quality indices to quantify various pollution levels. Because these air quality indices are not standardized,

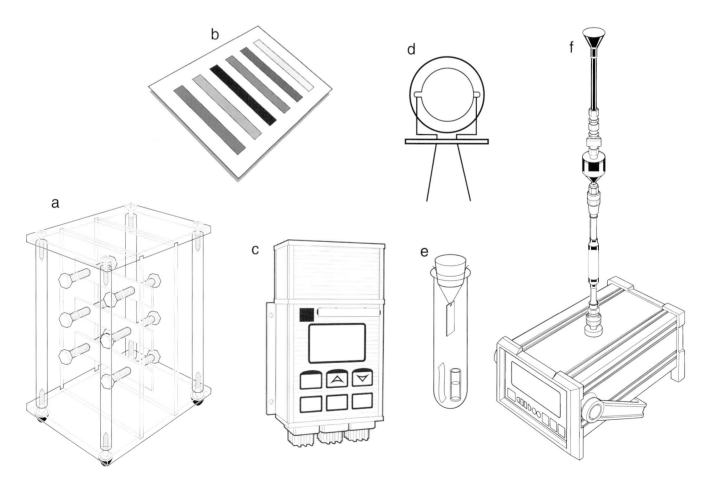

Figure 53. Semi-quantitative monitoring: (a) metal coupons; (b) dosimeter coated with egg tempera; (c) metal-coated piezoelectric quartz (OnGuard); (d) piezoelectric quartz sensor; (e) accelerated corrosion test (Oddy test); (f) real-time particulate monitor (quantitative).

a careful examination of the calculation methods is required (Hewings 2000). Fortunately, this situation will soon change as air quality indices in North America are currently undergoing a standardization process. Pollutant levels found at the closest monitoring station will most likely be different from those at the actual museum location; therefore, the stated level of pollutants provides only an estimation of the levels surrounding the museum. The indices include information on the levels of some key pollutants, which allows their levels to be estimated not only outside the museum but also inside by using the "100, 10, 1" rule (see section " '100, 10, 1' Rule for the Level of Outdoor Pollutants"). Indoor levels can also be estimated by referring to Appendix 4. Due to the high cost of extensive monitoring, many museums must be content with these approximate levels.

When doing a general assessment of a building's performance, monitoring of both outdoor and indoor pollutants should be considered. Nitrogen dioxide (NO_2) and fine particles ($PM_{2.5}$) are two good candidates for monitoring. The outdoor levels of NO_2 tend to remain high compared to sulphur dioxide and, indoors, NO_2 does not have the same capacity to be adsorbed by a material's surface. Its infiltration through the museum will be easy to track. Monitoring particles with a diameter less than 2.5 μm allows for the determination of the actual efficiency of the particle filter system and the barrier capacity of the building. Ozone can be added to the monitoring campaign where pertinent. There is usually not much ozone during the winter in Canada, but in the summer there are an increasing number of ozone alerts in the Quebec City – Windsor corridor. Hydrogen sulphide (H_2S) should be monitored if the building houses a precious metal collection. Because it has such a great effect on the deterioration rate (even if it is not the main pollutant), the level of water vapour expressed as a percentage of RH should be systematically included in all investigations.

The time at which the sampling is conducted can have a great impact on the results. For example, monitoring campaigns are often conducted in new buildings before the collection is moved in, and an empty room will always give higher levels of outdoor pollutants than a room full of collections (this difference is due to the presence or absence of high-sorbent materials in the room). Weather and human activities also influence the level of pollutants: ozone levels become high in the afternoon during sunny days, and pollutants due to car traffic increase

during rush hours — early in the morning and at the end of the afternoon. During busy periods (weekends and mid-afternoons are the busiest periods for many museums), the high numbers of visitors can release substantial amounts of water vapour, hydrogen sulphide, and ammonia. There are also seasonal fluctuations. The most conservative approach to monitoring is to sample during the worst-case scenarios when the highest levels are expected.

It is sometimes beneficial to know the daily fluctuations of some key pollutants as well as their average levels outside and inside the building over the course of a week. Monitoring daily fluctuations provides information such as how well the HVAC system controls the ozone cycles and what the impact of a high numbers of visitors is, while the weekly or monthly average levels of pollutants provide information about the overall environmental performance. Seasonal levels are also important, especially for water vapour.

The interior environment of enclosures may also need to be monitored. If the enclosure contains objects that are sensitive to acetic acid, this pollutant should be watched closely. Likewise if the enclosure contains a hypersensitive collection (Table 15) or an object of special interest or value, any pertinent pollutants should be monitored as well as the state of the collection or the specific object. Daily fluctuations in RH inside the enclosure require monitoring, but there is seldom a need to measure the daily fluctuations of other pollutants as it is their average levels and long-term trends that are more significant.

During a general investigation, an examination of the overall collection and some specific sensitive objects will yield important information on the general preservation performance of the building. Any damage that is identified should be reported. If it is necessary to determine the cause of the damage, this should be considered as a specific investigation.

SPECIFIC INVESTIGATION

A specific investigation responds to a particular concern (such as specific damage or a high level of pollutants), to the identification of a specific source of pollutants, or to the need to quantify any change in environmental performance. The typical types of damage suspected to be due to pollutants are listed in Table 25. Damage may also be caused by other agents of deterioration working alone or in combination with pollutants. Some types of damage reported in Table 25 are so well known that further measurements become meaningless. To avoid wasting time and money, specific investigations should be limited to situations where unusual damage has occurred while the environmental conditions have remained constant.

Monitoring is often requested after observing damage. However, it is uncertain whether or not monitoring will be able to provide any useful information about the cause. Sometimes, it is unclear when the damage occurred. The current environmental conditions may not be the same as the conditions that caused the damage. However, even if the level of pollutants that caused the damage cannot be found with confidence, the source can usually be traced. The nature of the (past) activities, products, and objects around the damaged object, and the nature of the damage itself can provide valuable clues. In some cases, a sample from the altered surface can be taken and sent for analysis. The results of the analysis may then help in identifying the major pollutant involved. Tables 1–4 should also be consulted for typical sources and levels of pollutants, and the reactions they may cause. While these reference tables and Table 25 may not provide enough information to identify the cause of the specific damage or concern, they can help to point out the most probable pollutants involved. As with a general investigation, the RH level should also be monitored.

For objects located in a room, damage due to pollutants originating outside will usually appear slowly (after a few years or even many decades).

The probability of damage due to other agents of deterioration is much higher. However, in a few cases, damage may occur within a few months due to a high level of new emissive products (including cleaning liquids), the presence of previously deposited pollutants (such as residues from cleaning or impregnated salt), or the placing of emissive collections in inadequately ventilated rooms. Inside an enclosure, damage is mainly caused by the emission of new products or objects and can happen within a few months. Mould growth can also happen easily inside uncontrolled moist and cool enclosures.

If it is necessary to identify the nature of the products or the presence of their harmful components, this can be done quantitatively and semi-quantitatively using various methods. For example, the presence of sulphur can be identified with the lead acetate test as described in Table 26. Some common methods for testing paints as well as other types of products are discussed in CCI Technical Bulletin No. 21 *Coatings for Display and Storage in Museums* (Tétreault 1999a).

Another aspect of monitoring is to verify or quantify physical performance, such as airtightness of enclosures or the efficiency of filters used in an HVAC system or inside an enclosure. A review of methods used to measure airtightness is given in Appendix 7, but there is as yet no standardized measurement to determine the efficiency of filters, especially gas filters. However, some possible ways to determine when they need replacement are covered in the section "Controls at Building/Room Level."

Table 21. Protocol for investigating air quality

Before investigation	• Ensure that the investigation can fulfil the expectation of the client given the time, resources, and technology available. • Explain the results to be delivered and provide possible future interpretations and actions.
Preliminary investigation	• Do an exploratory survey of the area to identify outside sources of pollutants (traffic, industries), activities in rooms, and nature of materials (products and objects) in enclosures (refer to Tables 1–4 and Appendix 4A). Estimate airtightness of the structures and identify sorbent materials and, if present, identify the HVAC parameters related to air quality (refer to Tables 9–11). For damage, consult Table 25 to try to establish a link between the actual damage observed on objects and the common problem scenarios, and get information on possible exposure time. An investigation based on good knowledge of the sources of pollutants, their typical levels and reactivities, and the physics and chemistry of the environment, combined with an examination of the most common damage reported, can often provide sufficient information to establish the level of performance or to resolve a specific problem to the satisfaction of the client without further monitoring.
Sampling	**General** • Do short-term measurements during the worst possible conditions: room full of visitors in early afternoon, enclosures closed for at least 1 day before sampling. Exploratory short-term monitoring at different places may be necessary to determine the need and best sites for a more comprehensive monitoring campaign. • In order of priority, measure levels of NO_2, particles, and acetic acid (for enclosures). It can be important to know the daily fluctuations of some key pollutants and the average levels outside and inside the building over a few weeks. If the monitoring focusses on specific collections, monitor the most harmful pollutants associated with them (refer to Tables 3 and 15). • As much as possible, measurement or sampling of pollutants should be done both outside and in rooms, and should be done simultaneously to allow for comparisons. Sampling over 1 week is preferable to minimize fluctuations or improve limit of detection. Pollutant levels measured with direct reading monitors should be recorded constantly or at least taken every 4 h. If relevant, consider redoing the pollutant monitoring both outside and in rooms for each season as well as for changes in the RH and temperature. **Outdoors** • Possible monitoring sites are close to the air intake of the HVAC system, and areas protected from rain and wind. To avoid particle deposition (if applicable), place the head of sampling devices face down. Monitor three sites and take triplicate samples or measurements in each. Record meteorological parameters during the monitoring (e.g. wind direction frequencies, wind velocity, last days of rain, RH, temperature, smog period, date). **Room** • For global assessment, choose rooms with different functions such as low access storage, popular exhibit, and entrance hall. Do sampling close to the objects of interest; avoid dead corners or high airflow areas. Take triplicate samples or measurements for each site. Identify major materials in the room (type of floor, shelf units, etc.) and record the nature of the collection and the activities in the room during the sampling, such as group visits, evening ceremonies, days when the museum is closed. **Enclosure** • To monitor the highest possible levels of inside-generated pollutants, the enclosure should be kept closed for at least 1 day before sampling. Do sampling close to the objects of interest and avoid dead corners. Take triplicate samples or measurements for each enclosure. Identify the nature of the collection and the products inside the enclosure.
Shipping and storage of samples	• Place samples in a container, then seal the container securely and identify it clearly. Follow manufacturer's instructions concerning maximum temperature for shipping and storage, and respect the maximum delay allowed before analysis.
Report	• The report should include the following elements: - Average levels of airborne pollutants with standard deviation for each site in $\mu g\ m^{-3}$ or ppb. - Sampling conditions (sampling date, time of the day, sampling time, sampling flow rate) and meteorological and topographical conditions (if applicable). - Sampling and analytical methods (including apparatus, calibration procedures, interferences) with references or with detailed procedures. Include limit of detection of the technique. - Date of analysis, analyst's name and address. - If requested, report the activities in the room during the sampling and identify the collection and products present in room as well as in the enclosure tested. - If mandated to assess the risk, consult Chapter 3.

Many parameters are optional or they can be applied only to certain sites or certain sampling methods.
For more details on protocol consult ASTM (2001a, b) and Kozdron (1995).

TABLE 22. TECHNIQUES FOR MONITORING AIR QUALITY BASED ON SAMPLING PERIOD

Sampling period	Sampling and detection[a]	Pollutants targeted	Sensitivity[b]	Comments, limitations
A few minutes to a few hours *Gives only current levels. The enclosure must be in steady-state equilibrium before the sampling (1–3 days).*	Solid-phase microextraction (SPME) with chromatographic analysis	Formaldehyde, formic and acetic acids, and soon most pollutants	High to very high	Constraints with enclosures[c]. Multi-analysis with portable GC-MS for in situ analysis will be useful.
	Air sample bags and chromatographic analysis	Most pollutants	Low to medium	Not commonly used. The sensitivity can be high if SPME of the air sample is made.
	Forced trapping sampler with chromatographic or spectrometry analysis	Inorganic pollutants	High to very high	The volume of air is pumped; not suitable in small enclosures.
	Forced diffusion through sampler, direct reading (such as short-term Dräger tubes)	Most pollutants	Low	Easy test[d]. The volume of air is pumped; not suitable in small enclosures. Only gives indication of high levels of formaldehyde, formic and acetic acids. Cannot differentiate the acids. Constraints with enclosures.
	Particle sampler, direct reading	Fine and coarse particles	Low to medium	Low limit of detection due to short sampling period. Constraints with enclosures.
8–24 h *Gives average or daily average levels. Enclosures must be in steady-state equilibrium (1–3 days) before sampling.*	Diffusion tubular sampler and chromatographic analysis	Most pollutants	Low	Low limit of detection due to short sampling period. Gives an indication of high levels of pollutants.
	Diffusion through sampler, direct reading (such as Dräger diffusion tubes)	Most pollutants	Low	Easy test. Only gives indication of high levels of formaldehyde, formic and acetic acids. Cannot differentiate the acids.
	Particle sampler, direct reading	Fine and coarse particles	High to very high	Constraints with enclosures.
	pH-sensitive dosimeter (pH strip)	Acid and alkaline compounds	Low to medium	Easy test. Gives indication of moderate to high levels of carboxylic acids. Cannot differentiate the acids. Not appropriate for long-term exposure due to sensitivities to oxidants.
1–3 weeks (or more) *Gives weekly (or more) average levels.*	Diffusion tubular sampler with chromatographic analysis	Most pollutants	High to very high	Gives average levels of the pollutants.
	Metal coupons (Ag or Cu) with electrochemical analysis	Corrosive pollutants and RH	High to very high	Not necessarily more sensitive than objects made of same metal. Can prevent damage to less sensitive objects.
	Metal coupons (Ag or Pb), direct observation	Corrosive pollutants and RH	Medium to high	Easy test. As above.
	Glossy or sticky surfaces with microscopic or reflectance measurement	Coarse and super coarse particles	Unknown	Low sensitivity to fine particles.
	Any well characterized materials with (usually) laboratory analysis	Agents of deterioration reacting with the material	Medium to very high	Not always an easy test. As for metal coupons with electrochemical analysis.
Continuous (direct reading) *Gives current levels, usually shows levels over time.*	Particle sampler	Fine and coarse particles	Low	Low limit of detection if data are not cumulative. Constraints with enclosures.
	Various technologies	Inorganic pollutants, formaldehyde, oxygen	Medium to very high	Limited mobility of the instrument. Constraints with enclosures. Need periodic calibration.
	Hygrometers	RH	Medium to very high	Easy test. Not cheap but a must for museums. Tends to be inaccurate in extremely low and high RH conditions. Needs periodic calibration.
	Piezoelectric crystal coated with metal (Ag or Cu) (OnGuard)	Corrosive pollutants and RH	High to very high	More sensitive than objects made of same metals.

a: Methods for most of the techniques are in Tables 23 and 24.

b: Sensitivity refers to pollutants at roughly the levels expected for a preservation target of 10 yrs (Table 5).

c: Refers to the problems of inserting the sampler while preserving the steady-state equilibrium within the enclosure, the small volume of an enclosure, making a hole, the need for electric wires, or the lifetime of the battery.

d: Easy test refers to low cost and the ability to obtain results without the need for outside expertise. All the other methods require an environmental professional for analysis in a laboratory or for an in situ investigation with special instruments.

TABLE 23. TECHNIQUES FOR SAMPLING AND DETECTING AIRBORNE POLLUTANTS

Airborne pollutants	Sampling techniques (exposure time)	Detection techniques	Limit of detection (μg m^{-3})[a]	References[b]
Acetic acid	SPME (30 min)	GC/MS	2	Van Bommel et al. (2001); Ryhl-Svendsen et al. (2002)
	Diffusion tubular sampler (14 days)	HPLC	44	Gibson et al. (1997a)
	Diffusion through sampler (8 h)	Graduated scale [c]	3100	Leichnitz (1989)
Ammonia	None; continuous airflow	FTIR spectroscopy	6	Griffith et al. (2000)
	Diffusion through sampler (8 h)	Graduated scale	1800	Leichnitz (1989)
Carbonyl sulphide	Diffusion tubular sampler (few weeks)	Fluorescence spectroscopy	0.3	Ankersmit et al. (2000)
Formaldehyde	None; continuous airflow	Fluorescence spectroscopy	0.1	Areo-Laser (nd)
	Diffusion (through porous tube) sampler (24 h)	HPLC	0.2	Uchiyam et al. (1999)
	SPME (10 min)	GC/FID	1	Koziel et al. (2001)
	None	Electronic resistance	10	Mazurkiewicz (2001); Environmental Sensors Co. (nd)
	Liquid reagent sampler (24 h)	Colorimetry	10	ASTM D5014 (2001c)
	Diffusion tubular sampler (7 days)	HPLC	130	Gibson et al. (2001)
	Diffusion sampler (8 h)	Graduated scale	400	K & M ChromAir (nd)
Formic acid	SPME (30 min)	GC/MS	7	Van Bommel et al. (2001); Ryhl-Svendsen et al.(2002)
	Diffusion tubular sampler (14 days)	HPLC	13	Gibson et al. (1997a)
	Forced diffusion through sampler (few minutes)	Graduated scale	1900	Leichnitz (1989)
Hydrogen peroxide	Forced diffusion tubular sampler (24 h)	Chemiluminescence	0.03	Strigbrand et al. (1996)
Hydrogen sulphide	Diffusion tubular sampler (few weeks)	Fluorescence spectroscopy	0.04	Ankersmit et al. (2000)
	None; continuous airflow	Electrical resistance	3	Arizona Instrument (nd)
	Diffusion through sampler	Graduated scale	1800	Leichnitz (1989)
Nitric acid	Forced diffusion tubular sampler; PLV [d] (24 h)	Ion chromatography	0.1	Marshall et al. (1992)
Nitrogen dioxide (usually include all NO$_x$)	Simple sampler (30 days)	Ion chromatography	0.06	Yamada et al. (1999)
	Forced diffusion tubular sampler; PLV (24 h)	Ion chromatography	0.1	De Santis et al. (1996)
	None; continuous airflow	Chemiluminescence	0.1	Aero-Laser (nd); Bernard et al. (1997)
	Diffusion tubular sampler (14 days)	Colorimetry	3	Bernard et al. (1997)
	Liquid reagent sampler; PLV (1 h)	Colorimetry	4	ASTM D1607 (2001d)
	Diffusion through sampler	Graduated scale	2400	Leichnitz (1989)
Ozone	None; continuous airflow	Chemiluminescence	0.2	Aero-Laser (nd)
	Diffusion tubular sampler (30 days)	Colorimetry	0.6	Bernard et al. (1999)
	Simple sampler (30 days)	Ion chromatography	1	Yamada et al. (1999)
	Diffusion sampler (8 h)	Graduated scale	160	K & M ChromAir (nd)
Particles	Cyclone separator and filters; PLV (24 h)	Gravimetry	0.05	Nazaroff et al. (1992); EPA (1998)
	Cyclone separator; PLV (1 min)	Photometric	0.1	MIE (2002)
Sulphur dioxide	Forced filter sampler; PLV (90 min)	Ion chromatography	0.1	Ferek et al. (1997)
	None; continuous airflow	Fluorescence spectroscopy	0.1	Ferek et al. (1997); Luke (1997)
	Diffusion sampler (30 days)	Ion chromatography	3	Yamada et al. (1999)
	Diffusion through sampler	Graduated scale	1700	Leichnitz (1989)

a: For most techniques, the limit of detection can be reduced by increasing the exposure period.

b: These references are not necessarily considered standard or even widely used; other valuable methods also exist.

c: Just read the graduated scale, no analysis required, cost is low.

d: PLV means pump large volume; accumulation of pollutants by air pumping is required. This method may not be appropriate for monitoring a small enclosure due to decreasing amounts of pollutants in the enclosure during the sampling.

TABLE 24. SEMI-QUANTITATIVE MONITORING METHODS

Airborne pollutants	Name	Sampling techniques (exposure time)	Detection techniques[a]	References
Specific pollutants				
Acetic acid from cellulose acetate films	A-D strips	pH-sensitive colorant (24 h)	Graduated colour scale	Image Permanence Institute (1997)
	Danchek film indicators	pH-sensitive colorant (24 h)	Graduated colour scale	Conservation by Design Limited (nd)
Groups of pollutants				
Acidic and alkaline compounds	A-D strips	pH-sensitive colorant (7 days)	Graduated colour scale	Nicholson and O'Loughlin (1996, 2000)
	pH strip with glycerol	pH-sensitive colorant (24 h)	Graduated colour scale	Tétreault (1992b, 1999a)
Particles	Dust deposition measurement	Dust deposition on glass slides or sticky pads (7 days or more)	Gloss surface reflectance	Adams et al. (2001); Eremin et al. (2000)
Pollutants contained in a product	Accelerated corrosion test (Oddy test)	Metal coupons (Ag, Cu, Pb); evaluate off-gassing of a product at 60°C, 100% RH (28 days)	Visual observation compared to blank samples	Green et al. (1995); Tétreault (1999a)
Environment (pollutants, radiation, RH, and temperature)				
Corrosive gases, RH	Metal coupons	Silver, copper, nickel, or zinc coupons (1–3 months)	Electrochemical analysis (corrosion film thickness measurement)	Purafil (1998b); Johansson et al. (1998)
	OnGuard	Piezoelectric quartz crystal coated with metal (Cu or Ag) (30–90 days)	Crystal frequency measurement	Purafil (1998a)
"Total environment" (various environmental agents of deterioration)	Organic-coated piezoelectric crystal[b]	Piezoelectric quartz crystals coated with egg tempera or varnish (direct reading)	Crystal frequency measurement	Odlyha et al. (2002)
	Colloidal silver film strip	Colloidal silver film strip (few weeks to few months)	Visual comparison with a control or density measurement	Weyde (1972); Wilhelm 1993
	Dosimeter coated with egg tempera[b]	Various pigments in egg tempera medium (9 months)	Reflectance spectrometry	Bacci et al. (2000)
	Glass sensor[b]	Potassium- and calcium-rich glass slides (3–12 months)	IR spectrometry	Pilz (2000); Leissner (1997); Roemich (2002)
	Paint-based dosimeters[b]	Pigmented and non-pigmentedwhole egg and mastic medium (9 months)	Mass spectrometry	Van den Brink et al. (1998)
	Any materials[b]	Any well-characterized materials such as acid-free papers, leather (few months)	Adapted to the material used	

a: Most methods do not have a lowest limit of detection or it is not provided with the methodology.

b: These materials are well characterized and are similar or the same nature as the objects to be preserved. Chemical analysis or physical tests will be done on the dosimeter to observe any early stages of deterioration caused by all the environmental agents of deterioration. These methods are still under development and are not yet widely available.

TABLE 25. TROUBLE-SHOOTING COMMON PROBLEMS RELATED TO AIRBORNE POLLUTANTS

Possible causes of damage related to airborne pollutants are in italic; most damage occurs very slowly and cannot be noticed with confidence without appropriate documentation; sometimes newly reported damage on an object is "stable" and was caused by a previous inappropriate context/environment or inappropriate treatment. Refer to Appendix 2 for some quantified data or for references.

	Common causes, comments, and possible monitoring
Corrosion on lead objects or lead alloy component of an object	**Causes:** *Acid-emissive products inside display cases, especially liquid products such as paints, adhesives, or silicone sealants, wood products, and possibly other organic objects in the enclosure (some ethnographic objects).* **Comments:** Wood products and paint inside enclosures do not usually provide safe conditions for lead especially in high relative humidity (>70% RH). **Possible monitoring**[a]: Identify acid-emissive products, and monitor acetic acids, formic acid, and RH in the enclosure.
Fast tarnishing of silver (within 2–3 months)	**Causes:** *Sulphur-based products such as sulphur vulcanized rubbers in an enclosure, other objects in the enclosure (fatty acids from organic ethnographic objects, pyrite or sulphate-reducing bacteria in impregnated objects)*, high RH, *high number of visitors in inadequately ventilated rooms.* **Comments:** Any sulphur-based materials in an enclosure have the potential to tarnish silver especially when the RH is high. **Possible monitoring**[a]: Identify possible sulphur-based materials, and monitor hydrogen sulphide, carbonyl sulphide, and RH.
Efflorescence on carbonate-based objects	**Causes:** Salt contamination of objects, high RH and large fluctuation of RH levels, *acid-emissive products inside display cases, especially liquid products such as paints, adhesives, and silicone sealants, and wood products.* **Comments:** This is a common problem when the object has been impregnated with salt either from contact with the ground or from salt having been included during its fabrication. Fluctuations in RH probably cause the efflorescence. Absorbed acetate (mainly) and formate compounds may initiate new efflorescence or contribute to a higher rate of formation due to their high solubility. **Possible monitoring** [a]: Monitor RH, temperature, salt contamination.
Crizzling (corrosion) on glasses	**Causes:** Slow process of hydration of alkali-rich glass (>10% soda content) is the most common deterioration process. In some cases, acetic and formic acids and possibly formaldehyde react with alkali glass to form a salt. However, salts based on carbonate, sulphate, and chloride are more common. **Comments:** The hydration process can cause loss of transparency, cracks, flaking, and spalling of areas of glass. Keep the RH below 60%.
Discoloration of colorants or materials	**Causes:** Light fading (photo-oxidation), thermal degradation, *other objects in the enclosure (organic ethnographic objects), outdoor airborne pollutants,* damage on object due to contact with other materials. **Comments:** Common forms of discoloration are yellowing, browning, and fading. **Possible monitoring**[a]: Monitor visible and UV radiation, RH, and temperature; identify nature of products in contact with the object; identify any sources of heat.
Cracks, deformation, or embrittled organic objects	**Causes:** Photo-oxidation, *outdoor airborne pollutants,* visible and UV radiation, high RH or large fluctuation of RH levels, thermal degradation. **Comments:** Examples of embrittlement are tears in textiles and in papers, and plastic objects or components turning into powder or shattering. **Possible monitoring**[a]: Monitor primary visible and UV radiation before considering ozone and nitrogen dioxide. Consider anoxic enclosures for the most valuable or vulnerable objects.
Stains or residues on objects	**Causes:** Damage due to contact with another material, fatty materials, internal pollutants such as plasticizers in plastics and adhesives, water damage, insect or mould activities. **Possible monitoring**[a,a]: Identify nature of object and/or products in contact, consider analysis of the altered compound.
Sticky objects	**Causes:** Internal pollutants such as plasticizers in PVC flexible objects; damage due to contact with another material; slow hydrolysis or oxidation of some plastics; polyethylene glycol impregnated waterlogged wood objects. **Possible monitoring**[a]: Identify nature of object and/or products in contact.
Blooms on objects or ghost image on protective glass	**Causes:** Deposition of amine compounds, transfer of salt by contact, or deposition of fatty acid on a surface from paint media very close to it (a few micrometres).
Dust deposition on objects or in the bottom of display cases	**Causes:** *Usually lint from carpets or from visitor clothing. Renovation activities/maintenance (generation or resuspension of dust) can also be a source.* **Comments:** Coarse particles are not necessarily harmful to objects; however, they may have an impact on their aesthetics. Deposited fine particles can be hard to remove. **Possible monitoring**[a]: Identify possible unusual sources of dust, such as from renovations, and isolate objects from the source.
Damage to products caused by an object	**Causes:** Damage by contact or *by airborne mode with mineral objects with acid components or fatty ethnographic objects.* **Comments:** Example of this is a label that becomes discoloured by being in the same enclosure as a pyrite specimen, which releases sulphur compounds. **Possible monitoring**[a]: Identify nature of object and/or products in contact.
Unusual smell	**Causes:** *Vinegar odour from black-and-white cellulose acetate (CA) film/microfilm or some tri-dimensional CA objects, new or deteriorated plastic products, deteriorated natural organic objects (ethnographic objects, books, etc.), mould in humid environment.* **Comments:** Smell does not necessarily mean significant risk of damage to objects. **Possible monitoring**[a]: Identify nature of object and products; monitor VOC, but keep focus mainly on key pollutants or the most probable one.

a: Take action according to the following steps: protect objects from the harmful environment; re-evaluate the control strategies (avoid, block, reduce); treat objects, if necessary, to limit further damage; and monitor after the improvements.

Table 26. Lead acetate test

Purpose of test	To determine the presence of sulphur. Products containing sulphur compounds are known to discolour, corrode, and weaken some objects.
Sample preparation	Liquid samples should be dried on aluminum foil for 1 day; solid samples such as a gasket or fabric need no preparation.
Reagents to be used	Lead acetate strip dispenser: BDH Chemicals, 350 Evans Avenue, Toronto ON M8Z 1K5, and in some drugstores. Hydrogen peroxide 3–10% solution: available at any local drugstore.
Comments	Wet the lead acetate test strip with one or two drops of clean water. Put the wet strip and a small test sample in a glass tube as shown to the right; the test strip and sample should not be in contact. Use a flame to pyrolyse the sample in the glass tube; tilt the tube horizontally during pyrolysis to make sure the dense smoke reaches the test strip. The test strip will turn brown in the presence of sulphur. Remove the brown test strip from the glass tube and add one drop of hydrogen peroxide. The presence of sulphur compounds in the sample will be confirmed if the brown test strip turns white. This test can produce strong irritant odours and should be conducted under a fume hood. 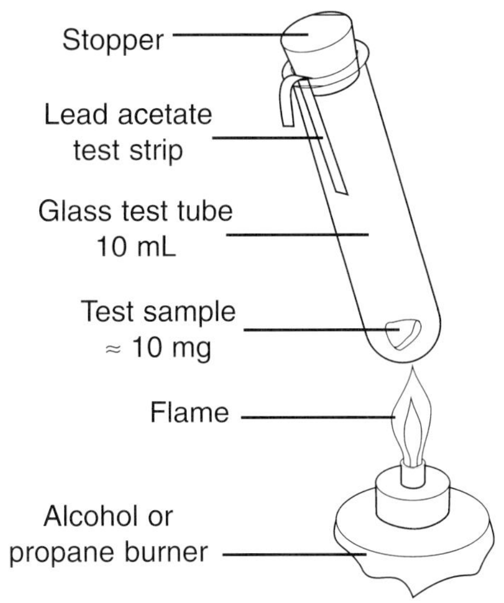

Stopper

Lead acetate test strip

Glass test tube 10 mL

Test sample ≈ 10 mg

Flame

Alcohol or propane burner

Figure 54. Schematic representation of the lead acetate test.

Acceptable	The product should not be used if the test strip turned white after being stained during pyrolysis. Products, such as vulcanized rubber, containing sulphur should not be used in contact with objects and should not share the same airtight or leaky enclosure. Alternative test: Azide test: Daniel and Ward (1982).

RECOMMENDED READING 7

Where an Internet version of the document exists, a link can be found on the CCI Web site (www.cci-icc.gc.ca/links/pollutants/index_e.shtml).

POLLUTION AND CONSERVATION

ASHRAE. "Museums, Libraries, and Archives." Chapter 21 in *Heating, Ventilating, and Air-Conditioning: Applications.* ASHRAE Handbook. Atlanta: 2003.

ASHRAE (American Society of Heating, Refrigerating and Air-Conditioning Engineers). "Air Contaminants." Chapter 12 in *Heating, Ventilating, and Air-Conditioning: Fundamentals.* ASHRAE Handbook. Atlanta: 2001.

ASHRAE. "Control of Gaseous Indoor Air Contaminants." Chapter 44 in *Heating, Ventilating, and Air-Conditioning: Applications.* ASHRAE Handbook. Atlanta: 1999.

Blades, N., T. Oreszczyn, B. Bordass, and M. Cassar. *Guidelines on Pollution Control in Museum Buildings.* London: Museums Association, 2000.

Feller, R.L. *Accelerated Ageing: Photochemical and Thermal Aspects.* Research in Conservation No. 4. Marina del Rey: The Getty Conservation Institute, 1994. <www.getty.edu/conservation/resources/reports.html>

Hatchfield, P.B. *Pollutants in the Museum Environment: Practical Strategies for Problem Solving in Design, Exhibition and Storage.* London: Archetype Publications, 2002.

Raphael, T., N. Davis, and K. Brookes. *Exhibit Conservation Guidelines: Incorporating Conservation into Exhibit Planning, Design and Fabrication.* CD-ROM. U.S. National Park Service, 1999.

Tétreault, J. *Coatings for Display and Storage in Museums.* CCI Technical Bulletin, No. 21. Ottawa: Canadian Conservation Institute, 1999.

UK-IIC (United Kingdom, International Institute of Conservation of Historic and Artistic Works). *Dirt and Pictures Separated.* Editorial board: S. Hackney, J. Townsend, and N. Eastaugh. London: The United Kingdom Institute of Conservation of Historic and Artistic Works, 1990.

RISK ASSESSMENT

American Chemical Society and Resource for the Future. *Understanding Risk Analysis: A Short Guide for Health, Safety, and Environmental Policy Making.* Internet edition (1998). <www.rff.org/misc_docs/risk_book.pdf>

Ashley-Smith, J. *Risk Assessment for Object Conservation.* Oxford: Butterworth Heinemann, 1999.

Cox, L.A. *Risk Analysis: Foundations, Models, and Methods.* International Series in Operations Research & Management Science 45. Boston: Kluwer Academic Publishers, 2001.

COST–BENEFIT ANALYSIS, ETHICS, AND RISK ASSESSMENT

CAC and CAPC (Canadian Association for Conservation of Cultural Property and the Canadian Association of Professional Conservators). *Code of Ethics and Guidance for Practices, third edition.* Ottawa: 2000. <www.cac-accr.ca/ehome.shtml>

Caple, C. *Conservation Skills: Judgement, Methods and Decision Making.* New York: Routledge, 2000.

Cassar, M. *Cost/Benefits Appraisals for Collection Care: A Practical Guide.* London: Museum & Galleries Commission, 1998.

Keene, S. *Managing Conservation in Museums, second edition.* Oxford: Butterworth-Heinemann, 2001.

Kopp, R.J., A.J. Krupnick, and M. Toman. *Cost-Benefit Analysis and Regulatory Reform: An Assessment of the Science and the Art.* Washington, DC: Resource for the Future, 1997. <www.rff.org/Documents/RFF-DP-97-19.pdf>

Lagas, R. *Cost-Benefit Analysis Guide For NIH IT Projects.* Office of the Deputy Chief Information Officer, 1999. <wwwoirm.nih.gov/itmra/cbaguide.html>

Northeast Document Conservation Center. "Planning and Prioritizing." Section 1, in *Preservation of Library & Archival Materials: A Manual* (edited by S. Ogden). Andover, MA: Northeast Document Conservation Center, 1999. <www.nedcc.org/plam3/manual.pdf>

Puglia, S. "Cost-Benefit Analysis for B/W Acetate: Cool/Cold Storage vs. Duplication." *Abbey Newsletter* 19 (1995). <palimpsest.stanford.edu/byorg/abbey/an/an19/an19-4/an19-410.html>

RECOMMENDED WEB SITES   8

These sites were accessed in December 2003; links can be found
on the CCI Web site <www.cci-icc.gc.ca/links/pollutants/index_e.shtml>.

POLLUTION AND CONSERVATION

Indoor Air Quality in Museums and Archives.

CONSERVATION

AINCA (Australian Network
for Information on Cellulose Acetate).
<www.nla.gov.au/anica/index.html>

Canadian Conservation Institute.
<www.cci-icc.gc.ca>

Conservation Information
Network (bibliographic database).
<www.bcin.ca>

GCI and IIC (The Getty Conservation Institute
and the International Institute for Conservation of
Historic and Artistic Works). Art and Archaeology
Technical Abstracts (bibliographic database).
<aata.getty.edu/NPS/>

Padfield, T. The Physics of the
Museum Environment.
<www.padfield.org/tim/cfys/>

Stanford University Libraries, Preservation
Department. Conservation Online (CoOL).

POLLUTION

APTI (Air Pollution Training Institute).
Basic Concepts in Environmental Sciences.
<www.epin.ncsu.edu/apti/ol_2000/home/
homefram.htm>

Atmosphere, Climate & Environment
Information Program, Encyclopaedia
of the Atmospheric Environment.
<www.doc.mmu.ac.uk/aric/eae/index.html>

EEA (European Environmental Agency).

EMEP (Co-operative Program for
Monitoring and Evaluation of the Long-Range
Transmission of Air Pollutants in Europe).
<www.emep.int/index_pollutants.html>

Environment Canada.
<www.mb.ec.gc.ca/index.en.html>

Environnement Québec.
<www.menv.gouv.qc.ca/air/inter_en.htm>

Minister of the Environment, Ontario.
<www.ene.gov.on.ca/air.htm>

University of Minnesota.
<www.me.umn.edu/courses/me5115/notes/>.

United States, EPA (Environmental
Protection Agency).
<www.epa.gov/ebtpages/air.html>

RISK ASSESSMENT
AND DECISION-MAKING

Mind Tools. Techniques for
Effective Decision Making.
<www.mindtools.com/pages/main/
newMN_TED.htm>

Risk World.

Society for Risk Analysis.

Figure 1. Corroded lead medals in an oak display case. Courtesy of the Musée du séminaire de Sherbrooke. (See p. 10 for more information.)

Figure 2. Tarnished silver-plated copper key ring. Continued cleaning of tarnish compounds will eventually remove the thin silver layer. (See p. 11 for more information.)

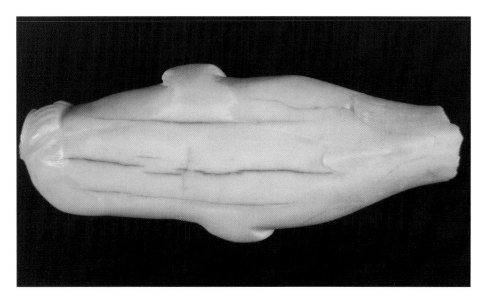

Figure 8. Inuit ivory sculpture with soot encrusted in the cracks. (See p. 15 for more information.)

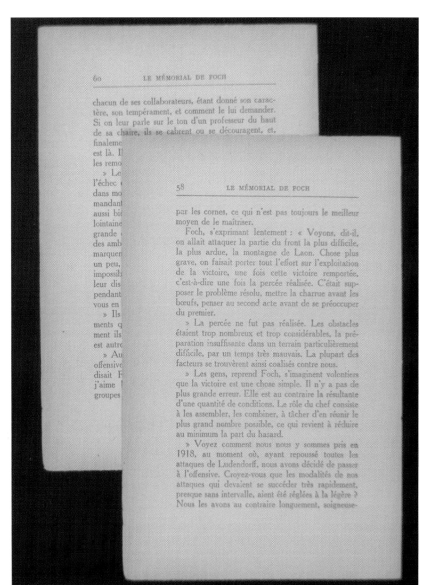

Figure 10. Browning of the edges of a page from a French book printed in 1929. The pollutants (mainly SO₂) were adsorbed on the edges and are slowly diffusing to the centre of the page. The pH of the darker zones is about 3.5 and the pH in the centre of the pages is 6.2. (See p. 16 for more information.)

Figure 11. Negative cellulose acetate sheet at advanced stage of degradation. The anti-curl layer and the emulsion layer have turned yellowish, and are 10% channelled and 100% blistered. The film base is yellowed and very brittle. It has completely degraded and converted from the ester to cellulose (i.e. regenerated cellulose). (See p. 16 for more information.)

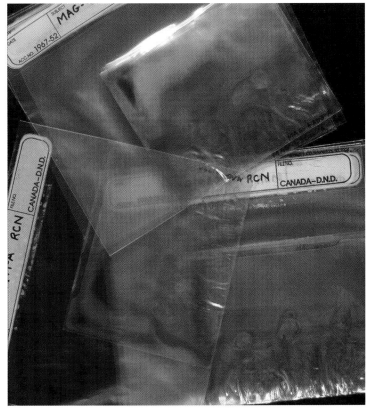

Figure 12. Negative cellulose nitrate sheets at advanced stage of degradation. The sheets have turned brown-yellow. (See p. 17 for more information.)

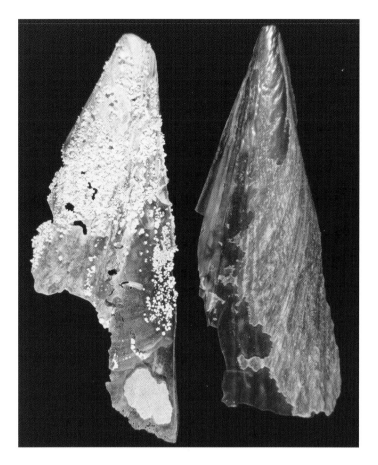

Figure 25. Efflorescence on seashells. The shell on the left has been exposed to acetic acid vapour and high RH. The shell on the right is the control sample. This type of deterioration is also known as Byne's disease in honour of Mr. Loftus St. George Byne who was the first to describe it in 1899 (Byne 1899). This type of efflorescence should not occur as long as RH is controlled, the sample is kept clean, and there are no materials nearby that emit high levels of acids. (See p. 29 for more information.)

Figure 29. Paper artwork stained by an acidic matboard after many years of contact. (See p. 40 for more information.)

Figure 47. A book covered with vegetable-tanned leather shows significant deterioration, known as red rot. This deterioration is caused by the transformation of sulphur dioxide into acid in the leather. The acidified leather surface chips off, and the underlying desiccated layer powders off when lightly rubbed. (See p. 66 for more information.)

Figure 48. Another example of red rot damage. The acidic vegetable-tanned leather turned black after being in contact with water. (See p. 66 for more information.)

Appendices: Data Tables

Appendix 1. Sources and Levels of Various Pollutants

Acetic acid (p. 99)	Formaldehyde (p. 101)	Nitric oxide (p. 102)	Particles, coarse (p. 104)
Amines (DEAE, ODA) (p. 100)	Formic acid (p. 101)	Nitrogen dioxide (p. 103)	Particles, fine (p. 104)
Ammonia (p. 100)	Hydrogen chloride (p. 102)	Oxygen (p. 103)	Peroxyacetyl nitrate (PAN) (p. 105)
Ammonium sulphate (p. 100)	Hydrogen peroxide (p. 102)	Ozone (p. 103)	Propionic acid (p. 105)
Carbon disulphide (p. 100)	Hydrogen sulphide (p. 102)	Particles, super coarse (p. 104)	Sulphur dioxide (p. 105)
Carbonyl sulphide (p. 100)	Nitric acid (p. 102)		

The levels of pollutants generated by materials (products or objects) are usually obtained using a test chamber. Some references are included even though no data are reported, but they describe the source and process of generation of the pollutant in the outdoor or indoor environment. X = data not reported, determined, or found. E3 or E6 = exponent 3 or 6.

Airborne pollutants/sources	Description	Concentration range, μg m^{-3}	References
Acetic acid			
Troposphere	Upper troposphere	0.3–0.9	Reiner et al. (1999)
Tropical forest	Wet and dry season range, daytime, 1999	0.9–9	Kuhn et al. (2001)
	Amazon, Brazil	1.7–4.5	Souza et al. (1999)
Urban sites		0.5–2.5	Graedel (1987a)
	Claremont, California, September 1985	9–18, peak at 24	Grosjean (1988)
	Cuidad Juarez, Chihuahua, Mexico, summer and winter	14, 35	Popp and Martin (1999)
	São Paulo, Brazil, July 1996	1.2–20	Souza et al. (1999)
Indoors	Storage and exhibition rooms	38–96	Grzywacz and Tennent (1994)
	Galleries of the Musical Instrument Museum (Brussels) and the Plantin-Moretus Museum (Antwerp), Belgium	82–106	Kontozova et al. (2002)
Adhesives	Different acrylics and PVAc	X	Down et al. (1996)
	Cardboard storage boxes containing PVAc	2600–4700	Dupont and Tétreault (2000b)
	Flooring products	X	Wilke et al. (2002)
Cellulose acetate film	Smelly silver gelatine sheet negative in a sealed jar for 24 h	150 000	Hollinshead et al. (1987)
	Smelly silver gelatine sheet negative with an air exchange of 10/day	920	Ryhl-Svendsen (2000)
Coatings	Oil-based paints, dried for 5 weeks, sampling made at equilibrium	20E3–65E3	Tétreault and Stamatopoulou (1997)
	Emulsion, moisture-cured urerhane or two-part epoxy, dried or cured for 5 weeks, sampling made at equilibrium	3000–22 000	Tétreault and Stamatopoulou (1997)
Linoleum	New samples	X	Jensen et al. (1993)
Medium density fibreboard (MDF)	Trade name: Sylvapan (Denmark)	220	Ryhl-Svendsen (2000)
Mould growth	Cultured mould	X	Bayer (1993)
Oak	Oak, solid wood, at least 15 yrs old	310	Ryhl-Svendsen (2000)
Silicone sealants	Bostik Glass 2680 silicone, acid type, cured for 3 days	14 000	Ryhl-Svendsen (2000)
	Bostik Glass 2680 silicone, acid type, cured for 7 days	880	Ryhl-Svendsen (2000)
	Bostik Glass 2680 silicone, acid type, cured for 29 days	94	Ryhl-Svendsen (2000)
	Bostik Industrial 2695 silicone, neutral type, cured for 27 h	<20	Ryhl-Svendsen (2000)
	Red oak, solid wood, about 10 yrs old, sampling made at equilibrium	6700	Tétreault and Stamatopoulou (1997)

Airborne pollutants/sources	Description	Concentration range, $\mu g\ m^{-3}$	References
Acetic acid (*continued*)			
Wooden enclosures	Various storage and display cabinets made of or having wooden panels	80–1800	Grzywacz and Tennent (1994)
	Various storage and display cabinets made of or having wooden panels	350–1800	Gibson et al. (1997)
	Various storage and display cabinets made of or having wooden panels	110–3200	Kontozova et al. (2002)
Amines, see also Ammonia			
Diethylamino ethanol (DEAE)			
HVAC system	Used as corrosion inhibitor	X	Volent and Baer (1985)
Octadecylamine (ODA)			
HVAC system	Used as corrosion inhibitor	About 1 in steam	Organ (1967)
Ammonia			
Clean troposphere		0.7	Seinfeld (1986)
Urban sites		7–18	Seinfeld (1986)
	Outside the Viking Ship Museum, Oslo	2	Dahlin et al. (1997)
Indoors	Library, gallery in a church, storage area and room in pulp-and-paper mill	0.6–15	Johansson et al. (1997)
	Exhibit and storage room in Japanese museum	16–40	Sano (2000)
	Newly finished building	20–60	Jarnstrom and Saarela (2002)
Human	Museum without HVAC system, 200 persons in museum	15	Dahlin et al. (1997)
	Museum without HVAC system, 400 persons in museum	22	Dahlin et al. (1997)
	Museum without HVAC system, 1000 persons in museum	30	Dahlin et al. (1997)
Concrete	Site built: emission rate: 30–70 $\mu g\ m^{-2}$ h	X	Jarnstrom and Saarela (2002)
Silicone sealants	Low odour type (i.e. Silicone GE II) or neutral type	X	Tétreault (1990)
Ammonium sulphate			
Outdoors	Formed by reaction with ammonium and sulphur compounds; same reaction may also happen indoors	X	Seinfeld and Pandis (1998)
Carbon disulphide			
Clear troposphere		0.05	Brimblecombe et al. (1992)
Mould growth	Cultured mould	X	Bayer (1993)
Urban sites		0.5	Brimblecombe et al. (1992)
Wool	Curtains in room exposed to the sunlight	0.35	Brimblecombe et al. (1992)
	Carpets, 900 g of wool in a room	0.03 (simulated)	Brimblecombe et al. (1992)
	Dry clothing, 20 persons (500 g of wool)	0.009 (simulated)	Brimblecombe et al. (1992)
	Wet clothing, 20 persons (500 g of wool)	0.035 (simulated)	Brimblecombe et al. (1992)
Carbonyl sulphide			
Clear troposphere		0.7	Brimblecombe et al. (1992)
		1.0–1.1	Graedel et al. (1981)
Urban sites		0.7	Brimblecombe et al. (1992)

Airborne pollutants/sources	Description	Concentration range, μg m^{-3}	References
Carbonyl sulphide (*continued*)			
Wool	Curtains in room exposed to the sunlight	1.4	Brimblecombe et al. (1992)
	Carpets, 900 g of wool in a room	0.12 (simulated)	Brimblecombe et al. (1992)
	Dry clothing, 20 persons (500 g of wool)	0.035 (simulated)	Brimblecombe et al. (1992)
	Wet clothing, 20 persons (500 g of wool)	0.14 (simulated)	Brimblecombe et al. (1992)
Formaldehyde			
Clean troposhere		0.5	Seinfeld (1986)
Urban sites	Claremont, California, September 1985	2.0–14	Grosjean (1988)
	Six cities in California, 1-yr average	6.3–7.1	Grosjean (1991)
		5–19	Graedel (1987a)
	Outside of office buildings and schools	14–26	Cavallo et al. (1993)
Indoors	Storage and exhibition rooms	11–46	Grzywacz and Tennent (1994)
	Office building and schools	13–70	Cavallo et al. (1993)
	Newly occupied building in Australia with HVAC, carpet, and furniture, avg. for first year	32–42	Dingle et al. (1993)
	Newly finished buildings (source: mainly floor covering)	13–37	Jarnstrom and Saarela (2002)
Urea formaldehyde-based wood products	Study made in the 1970s or 1980s	500–6000	Meyer and Hermanns (1986)
Wooden enclosures	Various storage and display cabinets made of or having wooden panels	50–470	Grzywacz and Tennent (1994)
Formic acid			
Troposphere	Upper troposphere	0.1–0.4	Reiner et al. (1999)
Tropical forest	Wet and dry season range, daytime, 1999	0.73–8.0	Kuhn et al. (2001)
Urban sites	Claremont, California, September 14–18, 1985	6.1–15, peak at 20	Grosjean (1988)
	Upland, California, avg. of fifty-five 24-h samples in a 1-yr period (September 1988–1989)	3.5	Grosjean (1991)
	São Paulo, Brazil, July 1996	1.1–18	Souza et al. (1999)
	Cuidad Juarez, Chihuahua, Mexico, winter	3.2	Popp and Martin (1999)
Indoors	Storage and exhibition rooms	<0.6–28	Grzywacz and Tennent (1994)
	Galleries of the Musical Instrument Museum (Brussels) and the Plantin-Moretus Museum (Antwerp), Belgium	21–25	Kontozova et al. (2002)
Wooden enclosures	Various storage and display cabinets made of or having wooden panels	2–120	Grzywacz and Tennent (1994)
	Coin collection drawer	500	Ryhl-Svendsen and Glastrup (2002)
	Storage and display cabinets made of or having wooden panels	<40–1600	Gibson et al. (1997)
	Various storage and display cabinets made of or having wooden panels	16–440	Kontozova et al. (2002)
Adhesives	Cardboard storage boxes containing PVAc	1300–2900	Dupont and Tétreault (2000b)

Airborne pollutants/sources	Description	Concentration range, $\mu g\ m^{-3}$	References
Hydrogen chloride			
Urban sites		0.76–3.0	Graedel (1987a)
Hydrogen peroxide			
Desert and grassland		0.14–4.2	Graedel (1984)
Urban sites		14–42	Graedel (1987a)
Hydrogen sulphide			
Clean troposphere		0.014–14	Graedel (1984)
		0.01–0.02	Watts (2000)
		0.007–0.07	Graedel et al. (1981)
		0.05–1.0	Brimblecombe et al. (1992)
	Outside four Dutch museums in coastal and agricultural areas	0.068–0.19	Ankersmit et al. (2000)
Urban sites		0.14–0.7	Graedel (1987a)
		0.02–7	Graedel et al. (1981)
	Outside four Dutch museums in urban and industrial areas	0.12–0.59	Ankersmit et al. (2000)
	Outside the Sainsbury Centre for Visual Art, UK	0.13	Camuffo et al. (2001)
	Two industrial sites (petrochemical and pulp-and-paper) in Quebec, Canada, annual avg. from 1990 to 1994	0.3–1.3	Ministère de l'Environnement et de la Faune (1997)
Indoors	Four Dutch museums in different areas: urban, industrial, coastal, and agricultural	0.03–1.4	Ankersmit et al. (2000)
	Sainsbury Centre for Visual Art, UK	0.15	Camuffo et al. (2001)
Human	Based on 2.8 mg per person per day in a room	16 (simulated)	Brimblecombe et al. (1992)
Sulphur-based paint components	Reported without precise data	X	Miles (1986)
Sulphur-contaminated artifacts	Reported without precise data	X	Green (1992)
Nitric acid			
Clean troposphere		0.04–0.6	Seinfeld (1986)
Urban sites	California	1–8	Salmon et al. (1990)
		3–30	Graedel (1987a)
		8–130	Seinfeld (1986)
	Outside eight institutions in California	7–26	Hisham and Grosjean (1991a)
Indoors	Inside eight institutions in California	2–16	Hisham and Grosjean (1991a)
Nitric oxide			
Urban sites	Claremont, California, Sepember 1985	2–40	Grosjean (1988)

Airborne pollutants/sources	Description	Concentration range, $\mu g\ m^{-3}$	References
Nitrogen dioxide			
Clean troposphere		0.2–1	Seinfeld (1986); Graedel (1984)
	Rural sites in USA, annual avg. level in 1990s	15–19	EPA (2001c)
	Rural sites in central and east European countries, 1997	2–23	Fiala et al. (2002)
Urban sites	Urban sites in USA, annual avg. level in 1990s, some sites are higher	42–48, 100	EPA (2001c)
		100–500	Seinfeld (1986)
	Canadian cities in 1990s, annual avg. range	32–42	EC (1999)
	Urban sites in central and east European countries, 1997	3–66	Fiala et al. (2002)
	Claremont, California, September 1985, includes all NO_x less NO	28–120, peak at 250	Grosjean (1988)
	Outside eight institutions in California	40–130	Hisham and Grosjean (1991a)
Indoors	Inside eight institutions in California	20–90	Hisham and Grosjean (1991a)
Cellulose nitrate film collection	In the room, value includes all NO_x	20	Erickson (1990)
	Cellulose negative films inside a plastic bag	3800	Hollinshead et al. (1987)
Oxygen			
Atmosphere		2.8E8 (20.95%)	
Ozone			
Clean troposphere		40–160	Seinfeld (1986)
		2–200	Graedel (1984)
	Rural sites in USA in 1990s, annual avg. range	200–220	EPA (2001b)
	Rural sites in central and east European countries, 1997	130–230	Fiala et al. (2002)
Urban sites	California, summer, avg. of 12 days of 24-h avg.	18–62	Hisham and Grosjean (1991a)
	California, daily range, July 12, 1984	10–350	Cass et al. (1989)
	Canadian cities in 1990s, annual avg. range	34–42	EC (1999)
	Urban sites in central and east European countries, 1997	130–290	Fiala et al. (2002)
	Urban and suburban sites in USA in 1990s, annual avg. range	200–240, max 400	EPA (2001b,c)
		200–1000	Seinfeld (1986)
Electrostatic precipitator	System running at 13 kV; measurement from the output less background level	50	Thomson (1965)
Photocopier	One machine in a sealed room, running a few hours	4–300	Claridge (1983)
Photocopier/laser printer	Three photocopiers and four laser printers in a room	<6	Schmith Etkin (1992)
Ozone generator	Run at a high setting in room with and without door open	200–600	EPA (2001a)
	Information from an HVAC contractor	40–80	Druzik (1999)
	In room (no further information)	<100	Bower (1991)
	Increases ambient level by up to 18 $\mu g\ m^{-3}$	ambient level +18	Bowser and Fugler (2002)

Airborne pollutants/sources	Description	Concentration range, μg m^{-3}	References
Particles, super coarse: > PM$_{10}$			
Human danders	Modern offices in Oslo, skin scales	X	Kruse et al. (2002)
Lints	Clothing and carpet fibres at indoor sites	X	Yoon and Brimblecombe (2000)
Particles, coarse: PM$_{10}$			
Clean troposphere	Rural sites in USA, annual avg. range in 1990s	19–24	EPA (2001b)
Urban sites	11 cities in Canada from 1984 to 1993, 24-h avg.	2–65	MOE (1999)
	Canadian cities in 1990s, annual avg. range	20–30	EC (1999)
	19 Canadian cities in 1993, avg. range per site	2–65	EC (1997)
	Birmingham, UK, July 1998	5–40	Jones et al. (2000)
	Urban sites in USA, annual avg. range in 1990s	25–31	EPA (2001b)
Concrete	Modern offices in Oslo	X	Kruse et al. (2002)
Particles, fine: PM$_{2.5}$			
Clean troposphere	Rural sites in USA, annual avg. level in 1990s	4–12, (in some sites: >20)	EPA (2001b)
	Rural sites in USA, annual avg. range in 1990s	15–30	EPA (2001c)
Urban sites	Urban sites in USA, annual avg. range in 1990s	7–40, some >65	EPA (2001c)
	11 cities in Canada from 1984 to 1993, 24-h avg.; 34% of the PM$_{2.5}$ is from fuel combustion	1–45	MOE (1999)
	Birmingham, UK, July 1998	5–15	Jones et al. (2000)
	Five museums in California, summer 1987	10–40	Nazaroff et al. (1993)
	Five museums in California, winter 1987	2–160 (peak)	Nazaroff et al. (1993)
	Canadian cities in 1990s, annual avg. range	10–13	EC (1999)
	19 Canadian cities in 1993, avg. range per site	2–40	EC (1997)
	Nine stations in Alberta, 24-h avg. for April to June 1998	12	Alberta Environmental Protection (1998)
	Forest fire at 300 km for the Calgary monitoring station, May 5, 1998	320	Alberta Environmental Protection (1998)
Indoors	Office building within Harvard University (dust deposited on the surface after 5 working days)	4.9–5.5	Kildeso et al. (1999)
	Office building within Harvard University N = 13AE/h, avg. time: 35 h	4.6–5.6	Kildeso et al. (1999)
	Office building with possible smokers N = 14AE/h, avg. time: 6–16 h	17–27	Kildeso et al. (1999)
	Office building with possible smokers N = 58AE/h, avg. time: 8 h	7.4–8.6	Kildeso et al. (1999)
Aerosol humidifiers	Museum: copper and salt was identified on objects	X	Rogers and Costain (1980)
Burning candles and incense	Literature survey	X	EPA (2001d)
Concrete	Modern offices in Oslo	X	Kruse et al. (2002)
Cooking	House in Birmingham, UK, no HVAC system, periodically open windows, July 1998, gas cooking	50	Jones et al. (2000)

Airborne pollutants/sources	Description	Concentration range, μg m^{-3}	References
Peroxyacetyl nitrate (PAN): a component of the photochemical smog			
Urban sites	Claremont, California, September 14–18, 1985	1–59	Grosjean (1988)
	Milan, Italy, February 10–21, 1993	0.5–16	De Santis et al. (1996)
Typical values		0.25–25	Allegrini et al. (1989)
Propionic acid			
Linoleum	40 days old	150	Jensen et al. (1993)
Sulphur dioxide			
Clean troposphere		3–30	Seinfeld (1986)
		0.13–0.32	Graedel et al. (1981)
	Rural sites in central and east European countries, 1997	2–37	Fiala et al. (2002)
	Rural site in USA, annual avg. level in 1990s	16–26	EPA (2001c)
Urban sites	Urban sites in USA, annual avg. level in 1990s	11–17, some peaks up to 400	EPA (2001c)
	Canadian cities in 1990s, annual avg. range	11–16	EC (1999)
	Urban sites in central and east European countries, 1997	6–94	Fiala et al. (2002)

Appendix 2. Damage to Various Materials due to Airborne Pollutants in Indoor Environments

Acetic acid (p. 106)	Formaldehyde (p. 109)	Oxygen (p. 119)
Amines (DEAE, ODA) (p. 108)	Formic acid (p. 110)	Ozone (p. 119)
Ammonia (p. 108)	Hydrogen peroxide (p. 110)	Ozone, nitrogen dioxide, and PAN (p. 122)
Ammonium sulphate (p. 108)	Hydrogen sulphide (p. 111)	Particles, fine (p. 123)
Anoxic environment (p. 108)	Nitric acid (p. 112)	Peroxyacetyl nitrate (PAN) (p. 123)
Carbonyl sulphide (p. 108)	Nitric oxide (p. 115)	Sulphur dioxide (p. 123)
Chlorine (p. 109)	Nitrogen dioxide (p. 115)	Sulphur dioxide and nitrogen dioxide (p. 125)
Fatty acids (p. 109)	Nitrous oxide (p. 119)	Water vapour (p. 127)

Abbreviations: AES = Auger electron spectroscopy (V); CI/IS = crystallinity index by infrared spectroscopy (%); CM = colour measurement (visible; CIE system: $\Delta E = 2$); CHR = chromatography technique; CR = contact resistance (%); DH = degree of hydrolysis; DP = degree of polymerization by viscometric determination (%); EA = electrochemical analysis (V); EDXA = energy-dispersive X-ray analysis (V); ET = ellipsometric technique (V); FE = folding endurance (%); FT/SEM = film thickness measurement with scanning electron microscopy (V); GYC = gypsum content (LD); pH = by aqueous extraction or on surface, NCD (%); M = modelled (V or %); PD = photographic densitometer (%); RM = reflectance measurement (% based usually on wavelength that causes the highest change); RTT= radiotracer technique, NCD (LD); SC = sulphate content measurement, NCD (LD); SD = silver density (%); TS = tensile strength (%); VO = visual observation (V); WG = weight gain measurement (LD); WL = weight loss measurement (LD); XRD = X-ray diffraction spectroscopy.

E3 or E6 = exponent 3 or 6; EL = estimated level from the original papers or from typical levels; LD = limit of detection of the technique; NCD = not necessarily considered as deterioration; PO = periodic observations up to; V = visible damage or 40 nm film thickness; WC = watercolour; X = data not reported, determined, or found; % = 5% of difference of the control.

Note: By default, exposures to pollutants are made at room temperature (20–25°C) and mainly in the dark. Grey zones identify the most sensitive objects to the pollutant at about 50% RH and room temperature. In a test chamber, the estimated outlet pollutant concentration was chosen rather than the inlet concentration. The quality of data can vary considerably. Many experiments were done with the exposition of pollutants at levels higher than those typically found in museums, or the quantification was rather suggestive or empiric. Not all data receive equal weight. This weight of evidence approach considers the quality and adequacy of the in situ observation or experience. Severe exposure conditions (i.e. high relative humidity, high temperature, or high levels of pollutants) are not considered primary data. High levels of pollutant in a short exposure may provide a misleading picture of the potential adverse effect, because some deterioration processes may induce effects that do not arise at lower levels. Some of these severe exposures have been kept in this Appendix to demonstrate their influence on the NOAEL and LOAED values. In some cases, where NOAEL and LOAEL are provided but there is uncertainty about the NOAEL value, a LOAED is also provided. Some references are included even though no data are reported; they provide a good description of the damage to materials caused by the pollutant in an indoor environment.

Airborne pollutants/objects	Methods	Exposure conditions	NOAEL–LOAEL, μg m^{-3}	LOAED, μg m^{-3} yr	References
Acetic acid (CH$_3$COOH)					
Calcium carbonate-based materials					
Bone	VO	54% RH, PO 1 yr	2E6–X	X	Brokerhof and Van Bommel (1996)
Eggshell (chicken)	VO	54% RH, PO 1 yr	50E3–87E3	30 000	Brokerhof and Van Bommel (1996)
Ivory	VO	54% RH, PO 1 yr	2E6–X	X	Brokerhof and Van Bommel (1996)
Shell	VO	54% RH, PO 1 yr, *Cypraea annulus*	12E3–50E3	6000	Brokerhof and Van Bommel (1996)
	VO	54% RH, PO 1 yr, *Loripes lacteus*	87E3–310E3	20 000	Brokerhof and Van Bommel (1996)
	VO	54% RH, PO 1 yr, *Chamelea striatula*	X–12E3	12 000	Brokerhof and Van Bommel (1996)

Airborne pollutants/objects	Methods	Exposure conditions	NOAEL–LOAEL, μg m^{-3}	LOAED, μg m^{-3} yr	References
Acetic acid (CH₃COOH) (*continued*)					
Cellulose					
Cotton	DP	45–50% RH, 786 days, Whatman papers in acidic box (also presence of other acid compounds)	4000–X	X	Dupont and Tétreault (2000a)
	DP	54% RH, PO 80 days, on Whatman papers	3000–20 000	3000	Dupont and Tétreault (2000b)
	pH	54% RH, 80 days, on Whatman papers	X–3000	5000	Dupont and Tétreault (2000b)
Colorant					
Basic fuchsin	CM	54% RH, PO 52 days on WC paper	7E3–25E3	3E6	Tétreault and Lai (2001)
Metals					
Brass	WL	100% RH, 30°C, 3 weeks	X–13 000	70	Clarke and Longhurst (1961)
Cadmium	WL	100% RH, 30°C, 3 weeks	130–1300	20	Clarke and Longhurst (1961)
	WL	100% RH, 30°C, 3 weeks	X–1300	10	Donovan and Stringer (1972)
	VO	Various acid-emissive products	X–X	X	Seabright and Trezek (1948)
	VO	Cadmium plates corroded in display case made of fibreboard and oak	X–X	X	Brown (1998)
Copper	WG	54% RH, PO 5 months	37 000–X	X	Tétreault (1992b)
	VO	50% RH, PO 120 days	1300–13 000	1100	Thickett (1997)
	VO	100% RH, PO 120 days	X–1300	90	Thickett (1997)
	WL	100% RH, 30°C, 3 weeks	130–1300	10	Clarke and Longhurst (1961)
	WL	100% RH, 30°C, 3 weeks	X–1300	10	Donovan and Stringer (1972)
	VO, EA	100% RH, 30°C, PO 3 weeks	X–25 000	70	López-Delgado et al. (1998)
Iron	WL	100% RH, 30°C, 3 weeks	X–1300	0.4	Donovan and Stringer (1972)
Lead	WG	34% RH, PO 6 months	860–2600	X	Tétreault et al. (1998)
	WG	23–44% RH, 76 days	X–1000	10	Eremin and Wilthew (1998)
	VO	50% RH, PO 4 months, oak sample in a jar	X–6700	300	Miles (1986); Tétreault and Stamatopoulou (1997)
	VO	50% RH, PO 120 days	X–1300	60	Thickett (1997)
	WG	54% RH, PO 1 yr	430–600	X	Tétreault et al. (1998)
	VO	PO 3.5 yrs, museum indoor environment (formaldehyde: 500 μg m^{-3})	320–X	X	Thicket et al. (1998)
	WG	75% RH, PO 6 months	520–860	80	Tétreault et al. (1998)
	WL	100% RH, 30°C, 3 weeks	X–1300	4	Donovan and Stringer (1972)
	WG	54% RH, PO 6 months, on tarnished lead	2600–12 000	3000	Tétreault et al. (1998)
	WG	75% RH, PO 6 months, on tarnished lead	860–2600	700	Tétreault et al. (1998)
Nickel	WL	100% RH, 30°C, 3 weeks	13E3–130E3	100	Donovan and Stringer (1972)
Silver	WL	100% RH, 30°C, 3 weeks	170E3–X	X	Donovan and Stringer (1972)
Steel	WL	100% RH, 30°C, 3 weeks	130–1300	0.8	Clarke and Longhurst (1961)
Zinc	WG	54% RH, PO 11 months	22E3–27E3	X	Tétreault (1992b)
	WL	100% RH, 30°C, 3 weeks	13–130	3	Clarke and Longhurst (1961)
	WL	100% RH, 30°C, 3 weeks	X–1300	0.6	Donovan and Stringer (1972)

Airborne pollutants/objects	Methods	Exposure conditions	NOAEL– LOAEL, $\mu g \ m^{-3}$	LOAED, $\mu g \ m^{-3}$ yr	References
Amines, see also Ammonia and Ammonium sulphate					
Diethylamino ethanol, DEAE ($(CH_3CH_2)_2N(CHCH_2OH)$)					
Linseed oil	CM	Indoor environment	X–14	10	Oshio (1992)
Varnished paintings	VO	2 months, indoor environment	X–50 (LD)	8	Biddle (1983)
	VO	Vapour release by highly concentrated solution	X–X	X	Williams (2001)
Octadecylamine, ODA ($CH_3(CH_2)_{16}CH_2NH_2$)					
Silver and copper	VO	Museum	X–X	X	Organ (1967)
Ammonia (NH_3)					
Cellulose nitrate	VO	Objects in open display and in storage boxes; compounds identified by FTIR	X–X	X	Tétreault and Williams 2002
Ammonium sulphate ($(NH_4)_2SO_4$), see also Sulphur dioxide and Particles					
Metals					
Aluminum	XRD	79% RH, 21 days, fine particles	540 $\mu g \ cm^{-2}$–X	X	Lobnig et al. (1996a)
Copper	XRD	93% RH, PO 10 h, fine particles	X–2 $\mu g \ cm^{-2}$	X	Unger et al. (1998)
	XRD	65% RH, 100°C, PO 5 days, fine particles	X–X	X	Lobnig et al. (1993
	XRD	75% RH, 100°C, PO 5 days, fine particles	X–X	X	Lobnig et al. (1993)
Zinc	XRD	60% RH, 21 days, fine particles	320 $\mu g \ cm^{-2}$–X	X	Lobnig et al. (1996b)
	XRD	65% RH, 21 days, fine particles	X–320 $\mu g \ cm^{-2}$	X	Lobnig et al. (1996b)
Natural resin varnishes	VO	Indoor environment, EL	X–55	X	Thomson (1986); Persson and Leygraf (1995)
Anoxic environment					
Colorants		Red lead becomes darker (photo-reduction*) when exposed to strong light intensities in a low oxygen environment	X–low oxygen	X	Maekawa (1998)
Carbonyl sulphide (COS)					
Metals					
Copper	EDXA	52% RH, PO 18 days	X–8700	30	Graedel et al. (1985)
	EDXA	80% RH, PO 5 days	X–27 000	3	Graedel et al. (1981)
	EDXA	95% RH, PO 3 days	X–8700	1	Graedel et al. (1985)
Silver	EDXA	4% RH, PO 54 days	X–8700	1000	Graedel et al. (1985)
	EDXA	70% RH, PO 54 days	X–8700	400	Graedel et al. (1985)
	EDXA	92% RH, PO 2 months	X–650	200	Franey et al. (1985)
	EDXA	95% RH, PO 14 days	X–8700	80	Graedel et al. (1985)

Airborne pollutants/objects	Methods	Exposure conditions	NOAEL– LOAEL, μg m^{-3}	LOAED, μg m^{-3} yr	References
Chlorine (Cl$_2$)					
Copper	WG	0% RH, 30°C, PO 30 h	X–59	0.008	Schubert (1988)
	EA	0% RH, PO 10 days	X–290	0.6	Fiaud and Guinement (1986)
	EA	75% RH, PO 10 days	X–290	0.6	Fiaud and Guinement (1986)
Fatty acids					
Glass frames and organic objects	VO	Indoor environment in enclosure causes bloom or ghost images	X–X	X	Williams (1989); Ordonez (1998)
Formaldehyde (CH$_2$O)					
Colorants	CM	44–52% RH, 12 weeks, watercolours studied: alizarin carmine, alizarin crimson, aurora yellow, basic fuchsin, brown madder, cadmium yellow, carmine, chrome yellow, copper phthalocyanine, crimson lake, curcumin, disperse blue 3, French ultramarine blue, gamboge, Hooker's green light, indigo, indigo carmine, mauve, new gamboge, pararosaniline base, Payne's grey, permanent magenta, permanent rose, Prussian blue, Prussian green, purple lake, rose carthame, rose dore, thioindigo violet, windsor yellow	150–X	X	Williams et al. (1992)
Glass	VO	Indoor environment, possibly caused also by acetic and formic acids	X–X	X	Riederer (1997); Oakley (1990); Ryan et al. (1993)
Metals					
Aluminum	WG	60±10% RH, 17–30°C, 6 months	6300–63 000	30 000	Duncan and Daniels (1986)
	FT/SEM	100% RH, 35°C, PO 30 days	5E6–X	X	Cermakova and Vlchkova (1966)
Brass	WG	60±10% RH, 17–30°C, 6 months	X–6300	3000	Duncan (1986)
Copper	VO	50% RH, PO 120 days	630–6300	800	Thickett (1997)
	WG	60±10% RH, 17–30°C, 6 months	63 000–X	X	Duncan (1986)
	VO	100% RH, PO 120 days	X–630	50	Thickett (1997)
	FT/SEM	100% RH, 35°C, PO 30 days	5E6–X	X	Cermakova and Vlchkova (1966)
Iron	FT/SEM	100% RH, 35°C, PO 30 days	X–30 000	100	Cermakova and Vlchkova (1966)
Lead	VO	50% RH, PO 120 days	6300–X	X	Thickett (1997)
	VO	100% RH, PO 120 days	X–630	30	Thickett (1997)
Silver	WG	60±10% RH, 17–30°C, 6 months	X–630	2000	Duncan (1986)
Zinc	WG	60±10% RH, 17–30°C, 6 months	6300–63 000	10 000	Duncan (1986)
	FT/SEM	100% RH, 35°C, 21 days	5000–10 000	200	Cermakova and Vlchkova (1966)
	FT/SEM	100% RH, 35°C, PO 30 days	X–30 000	1	Cermakova and Vlchkova (1966)

Airborne pollutants/objects	Methods	Exposure conditions	NOAEL–LOAEL, μg m^{-3}	LOAED, μg m^{-3} yr	References
Formic acid (HCOOH)					
Glass	VO	Indoor environment, possibly caused also by formaldehyde and acetic acid	X–X	X	Riederer (1997); Oakley (1990); Ryan et al. (1993)
Metals					
Cadmium	WL	100% RH, 30°C, 3 weeks	X–1100	8	Donovan and Stringer (1972)
	WL	100% RH, 30°C, 3 weeks	X–3800	100	Donovan and Stringer (1965)
Copper	VO	50% RH, PO 120 days	960–9600	1000	Thickett (1997)
	WG	54% RH, PO 135 days	8000–11 000	X	Tétreault et al. (2003)
	WG	75% RH, PO 135 days	4000–11 000	X	Tétreault et al. (2003)
	VO	100% RH, PO 120 days	X–960	40	Thickett (1997)
	VO, EA	100% RH, 30°C, PO 21 days	X–19 000	200	Bastidas et al. (2000)
	FT/SEM	100% RH, 35°C, PO 30 days	X–30 000	40	Cermakova and Vlchkova (1966)
Iron	WL	100% RH, 30°C, 3 weeks	X–1100	0.2	Donovan and Stringer (1972)
	FT/SEM	100% RH, 35°C, PO 30 days	X–30 000	3	Cermakova and Vlchkova (1966)
Lead	VO	50% RH, PO 120 days	X–960	100	Thickett (1997)
	WG	54% RH, PO 135 days	170–760	X	Tétreault et al. (2003)
	WG	75% RH, PO 135 days	760–2000	X	Tétreault et al. (2003)
	VO	100% RH, PO 120 days	X–960	40	Thickett (1997)
	WL	100% RH, 30°C, 3 weeks	X–1100	30	Donovan and Stringer (1972)
Lead tarnished 5–7 yrs	WG	54% RH, 33 days, source: particleboard	330–X	X	Tétreault et al. (2003)
	WG	75% RH, 33 days, source: particleboard	X–330	20	Tétreault et al. (2003)
	WG	100% RH, 33 days, source: particleboard	X–330	3	Tétreault et al. (2003)
Nickel	WL	100% RH, 30°C, 3 weeks	1100–11 000	30	Donovan and Stringer (1972)
Silver	WL	100% RH, 30°C, 3 weeks	130E3–X	X	Donovan and Stringer (1972)
Steel	WL	100% RH, 30°C, 3 weeks	3800–13 000	200	Donovan and Stringer (1965)
	WL	100% RH, 30°C, 3 weeks	13E3–48E3	1000	Donovan and Stringer (1965)
Zinc	WL	100% RH, 30°C, 3 weeks	X–1100	3	Donovan and Stringer (1972)
	WL	100% RH, 30°C, 3 weeks	X–3800	100	Donovan and Stringer (1965)
	FT/SEM	100% RH, 35°C, PO 30 days	X–30 000	1	Cermakova and Vlchkova (1966)
Hydrogen peroxide (HOOH)					
Silver image	VO	50% RH, 7 days, alkyd paints in indoor environment	X–42	0.8	Feldman (1981)
	CM	30% RH, 18 h, on polyester microfilm	X–1.4E6	1000	Adelstein et al. (1991)
	CM	50% RH, 18 h, on polyester microfilm	X–1.4E6	100	Adelstein et al. (1991)

Airborne pollutants/objects	Methods	Exposure conditions	NOAEL–LOAEL, $\mu g\ m^{-3}$	LOAED, $\mu g\ m^{-3}$ yr	References
Hydrogen peroxide (HOOH) (*continued*)					
	CM	82% RH, 18 h, on polyester microfilm	X–1.4E6	40	Adelstein et al. (1991)
	CM	82% RH, 18 h, 1950s non-iodide films	X–1.4E6	300	Adelstein et al. (1991)
	CM	82% RH, 18 h, on polyester microfilm	X–1.4E6	400	Adelstein et al. (1991)
	CM	82% RH, 18 h, on gold-treated polyester microfilm	1.4E6–X	X	Adelstein et al. (1991)
Hydrogen sulphide (H_2S)					
Metals					
Copper	EDXA	30% RH, PO 7 days	X–5000	3	Graedel et al. (1985)
	EDXA	39% RH, PO 7 h	X–8400	3	Franey et al. (1980)
	EA	75% RH, PO 10 days	X–28	0.9	Fiaud and Guinement (1986)
	EA	75% RH, PO 10 days, plus 191 $\mu g\ m^{-3}$ NO_2	X–28	0.5	Fiaud and Guinement (1986)
	EA	75% RH, PO 10 days	X–140	0.9	Fiaud and Guinement (1986)
	EA	75% RH, PO 10 days, only with 290 $\mu g\ m^{-3}$ Cl_2	X–140	0.6	Fiaud and Guinement (1986)
	EA	75% RH, PO 10 days, plus 290 $\mu g\ m^{-3}$ Cl_2	X–140	1	Fiaud and Guinement (1986)
	EDXA	85% RH, PO 7 days	X–21	2	Franey et al. (1980)
	EDXA	95% RH, PO 13 days	X–5000	0.1	Graedel et al. (1985)
	EDXA	93% RH, PO 2 days, H_2S only	X–3300	2	Franey (1988)
	EDXA	93% RH, PO 2 days with 400 $\mu g\ m^{-3}$ O_3	X–3300	1	Franey (1988)
	EDXA	93% RH, PO 2 days with 400 $\mu g\ m^{-3}$ O_3 and sunlight	X–3300	1	Franey (1988)
	EDXA	93% RH, PO 2 days, H_2S only	X–3300	2	Graedel et al. (1984)
	EDXA	93% RH, PO 2 days with 340 $\mu g\ m^{-3}$ O_3	X–3300	1	Graedel et al. (1984)
	EDXA	93% RH, PO 2 days with light (one solar constant)	X–3300	1	Graedel et al. (1984)
	EDXA	93% RH, PO 2 days with 400 $\mu g\ m^{-3}$ O_3 and light	X–3000	1	Graedel et al. (1984)
	WG	0–52% RH, 70 days (faster tarnishing rate)	X–57E6	70	Backlund et al. (1966)
	WG	52% RH, 70 days	X–57E6	70	Backlund et al. (1966)
	WG	75% RH, 70 days	X–57E6	7	Backlund et al. (1966)
	WG	100% RH, 70 days	X–57E6	2	Backlund et al. (1966)
Silver	ET	Indoor environment, dry room, 60 days	X–0.28	4	Bennett et al. (1969)
	WG	Dry and 75% RH, PO 21 days, with or without 5300 $\mu g\ m^{-3}$ SO_2	X–28	0.01	Pope et al. (1968)
	EDXA	5% RH, PO 8 days	X–5000	600	Graedel et al. (1985)
	EDXA	31% RH, PO 60 days	X–5000	600	Graedel et al. (1985)
	ET	Indoor environment, PO 8 days	X–0.28	0.1	Bennett et al. (1969)
	VO	Indoor environment, PO 6 months	X–0.57	0.3	Watts (1999)
	CM, RM	50% RH, PO 21 days	X–400	1	Ankersmit et al. (2000)
	EDXA	92% RH, PO 130 days	X–4500	200	Franey et al. (1985)
	EDXA	95% RH, PO 15 days	X–5000	70	Graedel et al. (1985)
	CR	30% RH, 30°C, PO 12 days, plus 2900 $\mu g\ m^{-3}$ Cl_2 and 2700 $\mu g\ m^{-3}$ SO_2	X–1400	8	Lorenzen (1971)

Airborne pollutants/objects	Methods	Exposure conditions	NOAEL–LOAEL, $\mu g \ m^{-3}$	LOAED, $\mu g \ m^{-3}$ yr	References
Hydrogen sulphide (H_2S) (*continued*)					
	CR	50% RH, 30°C, PO 12 days, plus 2900 $\mu g \ m^{-3}$ Cl_2 and 2700 $\mu g \ m^{-3}$ SO_2	X–1400	0.4	Lorenzen (1971)
	EA	75% RH, PO 10 days	X–28	1	Fiaud and Guinement (1986)
	EA	75% RH, PO 10 days	X–710	1	Fiaud and Guinement (1986)
	EA	75% RH, PO 10 days	X–140	2	Fiaud and Guinement (1986)
	EA	75% RH, PO 10 days, plus 3800 $\mu g \ m^{-3}$ NO_2	X–140	2	Fiaud and Guinement (1986)
	EA	75% RH, PO 10 days, plus 290 $\mu g \ m^{-3}$ Cl_2	X–140	1	Fiaud and Guinement (1986)
	CR	85% RH, 30°C, PO 2.5 days	X–170	0.7	Lorenzen (1971)
	CR	85% RH, 30°C, PO 2.5 days	X–1400	0.7	Lorenzen (1971)
	CR	85% RH, 30°C, PO 2.5 days, plus 2700 $\mu g \ m^{-3}$ SO_2	X–1400	0.5	Lorenzen (1971)
	CR	90% RH, 40°C, PO 2.5 days, plus 150 $\mu g \ m^{-3}$ Cl_2 and 1300 $\mu g \ m^{-3}$ SO_2	X–500	0.2	Lorenzen (1971)
	CR	85% RH, 30°C, PO 2.5 days, plus 2900 $\mu g \ m^{-3}$ Cl_2	X–1400	0.07	Lorenzen (1971)
Silver–copper alloy	EA	Dry air, 50°C, PO 16 h (Ag: 90%–Cu 10%)	X–1400	0.4	Simon et al. (1980)
	EA	35% RH, 50°C, PO 16 h	X–1400	0.4	Simon et al. (1980)
Micrograph film dyes					
Magenta	PD	50% RH, 50°C, PO 5 weeks, dye on Ilfochrome colour micrograph film CMM	7100–X	X	Zinn et al. (1994)
Yellow	PD	50% RH, 50°C, PO 5 weeks, dye on Eastman 5272, colour negative motion-picture film	7100–X	X	Zinn et al. (1994)
	PD	50% RH, 50°C, PO 5 weeks, dye on Eastman 52384, colour positive motion-picture film	7100–X	X	Zinn et al. (1994)
Lead white pigment	VO	31–84%, PO 7 days, on cotton paper	X–X	X	Hoevel (1985)
Silver image	PD	50% RH, 50°C, PO 5 weeks, black-and-white micrograph films	7100–X	X	Zinn et al. (1994)
Nitric acid (HNO_3)					
Colorants: top five most sensitive					
Litmus	CM	50% RH, PO 12 weeks, on WC paper	X–31	0.09	Salmon and Cass (1993)
Pararosaniline base	CM	50% RH, PO 12 weeks, on WC paper	X–31	0.09	Salmon and Cass (1993)
Aigami (malonyl-awobanin)	CM	50% RH, PO 12 weeks, on WC paper	X–31	0.3	Salmon and Cass (1993)
Alizarin crimson	CM	50% RH, PO 12 weeks, on WC paper	X–31	0.3	Salmon and Cass (1993)
Curcumin	CM	50% RH, PO 12 weeks, on WC paper	X–31	0.3	Salmon and Cass (1993)

Airborne pollutants/objects	Methods	Exposure conditions	NOAEL–LOAEL, μg m^{-3}	LOAED, μg m^{-3} yr	References
Nitric acid (HNO$_3$) (*continued*)					
Colorants: traditional natural organic					
Aigami (malonyl-awobanin)	CM	50% RH, PO 12 weeks, on WC paper	X–31	0.3	Salmon and Cass (1993)
Bitumen	CM	50% RH, 12 weeks, on WC paper	31–X	X	Salmon and Cass (1993)
Cochineal lake	CM	50% RH, 12 weeks, on WC paper	X–31	7	Salmon and Cass (1993)
Curcumin	CM	50% RH, PO 12 weeks, on WC paper	X–31	0.3	Salmon and Cass (1993)
Disperse blue 3	CM	50% RH, PO 12 weeks, on cellulose acetate	X–31	2	Salmon and Cass (1993)
Dragon's blood	CM	50% RH, PO 12 weeks, on WC paper	X–31	2	Salmon and Cass (1993)
Gamboge	CM	50% RH, 12 weeks, on WC paper	X–31	4	Salmon and Cass (1993)
Indian yellow	CM	50% RH, PO 12 weeks, on WC paper	X–31	3	Salmon and Cass (1993)
Indigo, synthetic	CM	50% RH, PO 12 weeks, on WC paper	31–X	X	Salmon and Cass (1993)
Indigo, natural blue 1	CM	50% RH, PO 12 weeks, on WC paper	X–100	0.6	Salmon and Cass (1993)
Lac lake	CM	50% RH, PO 12 weeks, on WC paper	X–31	2	Salmon and Cass (1993)
Litmus	CM	50% RH, PO 12 weeks, on WC paper	X–31	0.09	Salmon and Cass (1993)
Madder lake	CM	50% RH, 12 weeks, on WC paper	31–X	X	Salmon and Cass (1993)
Persian berries lake	CM	50% RH, PO 12 weeks, on WC paper	X–31	1	Salmon and Cass (1993)
Quercitron lake	CM	50% RH, PO 12 weeks, on WC paper	X–31	2	Salmon and Cass (1993)
Saffron	CM	50% RH, 12 weeks, on WC paper	31–X	X	Salmon and Cass (1993)
Sepia	CM	50% RH, 12 weeks, on WC paper	31–X	X	Salmon and Cass (1993)
Van Dyke brown	CM	50% RH, 12 weeks, on WC paper	31–X	X	Salmon and Cass (1993)
Weld lake	CM	50% RH, 12 weeks, on WC paper	X–31	5	Salmon and Cass (1993)
Colorants: synthetic organic colorants					
Acridone	CM	50% RH, 2 weeks, on WC paper	31–X	X	Salmon and Cass (1993)
Arylide yellow 10G	CM	50% RH, PO 12 weeks, on WC paper	X–31	0.6	Salmon and Cass (1993)
Alizarin crimson	CM	50% RH, PO 12 weeks, on Whatman paper	X–31	4	Salmon and Cass (1993)
Alizarin crimson	CM	50% RH, PO 12 weeks, on WC paper	X–31	0.3	Salmon and Cass (1993)
Aniline black	CM	50% RH, 12 weeks, on WC paper	31–X	X	Salmon and Cass (1993)
Basic fuchsin	CM	50% RH, PO 12 weeks, on WC paper	X–31	2	Salmon and Cass (1993)
Bright red	CM	50% RH, PO 12 weeks, on WC paper	31–X	X	Salmon and Cass (1993)
Dioxazine purple	CM	50% RH, 12 weeks, on WC paper	31–X	X	Salmon and Cass (1993)
Indigo syn.	CM	50% RH, PO 12 weeks (Indigotin, vat blue1)	31–X	X	Salmon and Cass (1993)
Indigo syn.	CM	50% RH, PO 12 weeks (Indigotin, vat blue1)	X–100	0.6	Salmon and Cass (1993)
Paliogen blue	CM	50% RH, 12 weeks, on WC paper	31–X	X	Salmon and Cass (1993)
Paliogen yellow	CM	50% RH, 12 weeks, on WC paper	31–X	X	Salmon and Cass (1993)
Pararosaniline base	CM	50% RH, PO 12 weeks, on WC paper	X–31	0.09	Salmon and Cass (1993)
Permanent magenta	CM	50% RH, 12 weeks, on WC paper	31–X	X	Salmon and Cass (1993)

Airborne pollutants/objects	Methods	Exposure conditions	NOAEL–LOAEL, μg m^{-3}	LOAED, μg m^{-3} yr	References
Nitric acid (HNO$_3$) (*continued*)					
Phthalocyanine blue	CM	50% RH, 12 weeks, on WC paper	31–X	X	Salmon and Cass (1993)
Phthalocyanine green	CM	50% RH, 12 weeks, on WC paper	31–X	X	Salmon and Cass (1993)
Prussian blue	CM	50% RH, PO 12 weeks, on WC paper	X–31	10	Salmon and Cass (1993)
Mauve	CM	50% RH, PO12 weeks, on WC paper	X–31	3	Salmon and Cass (1993)
Naphthol	CM	50% RH, 12 weeks, on WC paper	31–X	X	Salmon and Cass (1993)
Quinacridone red	CM	50% RH, 12 weeks, on WC paper	31–X	X	Salmon and Cass (1993)
Rose carthame	CM	50% RH, 12 weeks, on WC paper,	31–X	X	Salmon and Cass (1993)
Scarlet lake	CM	50% RH, 12 weeks, on WC paper	31–X	X	Salmon and Cass (1993)
Thioindigo violet	CM	50% RH, 12 weeks, on WC paper	31–X	X	Salmon and Cass (1993)
Toluidine red	CM	50% RH, 12 weeks, on WC paper	31–X	X	Salmon and Cass (1993)
Colorants: inorganic colorants					
Chrome yellow	CM	50% RH, PO 12 weeks, on Whatman paper	31–X	X	Salmon and Cass (1993)
Aureolin, cadmium yellow, manganese violet, vermilion	CM	50% RH, PO 12 weeks, on Whatman paper	31–X	X	Salmon and Cass (1993)
Orpiment	CM	50% RH, PO 12 weeks, on Whatman paper	X–31	5	Salmon and Cass (1993)
Orpiment, faded	CM	50% RH, PO 12 weeks, previously exposed to 7 μg m^{-3} yr	X–100	40	Salmon and Cass (1993)
Realgar	CM	50% RH, PO 12 weeks, on Whatman paper	X–31	10	Salmon and Cass (1993)
Colorants: inorganic colorants: inks					
Iron ink (gallotannate)	CM	50% RH, PO 12 weeks, on WC paper	X–31	3	Salmon and Cass (1993)
Iron ink (tannate)	CM	50% RH, PO 12 weeks, on WC paper	X–31	2	Salmon and Cass (1993)
Colorants: traditional Japanese colorants on silk (with or without mordants)					
Beni (orange and red), Kariyasu, Ai + Enju, Akane, Ukon, Ai + Kariyasu	CM	50% RH, PO 12 weeks	X–31	2	Salmon and Cass (1993)
Beni (orange), faded	CM	50% RH, PO 12 weeks, previously exposed to 7 μg m^{-3} yr	X–100	1	Salmon and Cass (1993)
Yama momo	CM	50% RH, PO 12 weeks	X–31	7	Salmon and Cass (1993)
Enju (rutin)	CM	50% RH, PO 12 weeks	X–31	0.5	Salmon and Cass (1993)
Kihada, Kuchi nashi, Woren, Yamahaji, Zumi, Shio, Seiyo akane, Enji, Shiko, Suo, Ai, Shikon, Ai + Kihada	CM	50% RH, 12 weeks	31–X	X	Salmon and Cass (1993)

Airborne pollutants/objects	Methods	Exposure conditions	NOAEL– LOAEL, $\mu g\ m^{-3}$	LOAED, $\mu g\ m^{-3}$ yr	References
Nitric oxide (NO)					
Silver image	SD	50% RH, 30 h, on triacetate support	420E3–X	X	Carroll and Calhoun (1955)
Nitrogen dioxide (NO$_2$)					
Calcium carbonate-based materials					
Limestone, marble, travertine	WG	90% RH, PO 6 weeks	4300–X	X	Johansson et al. (1988)
Colorants: top 4 most sensitive					
Curcumin	CM	90% RH, PO 10 days, on silk and on cotton	X–1900	1	Saito et al. (1993)
Enju (rutin)	CM	65% RH, PO 10 days, mordant Al, on cotton	X–1900	1	Saito et al. (1994)
Hematoxylin	CM	90% RH, PO 10 days, on cotton	X–1900	1	Saito et al. (1993)
Quercitron	CM	Indoor environment, PO 10 days, on cotton and on silk	X–70	1	Saito et al. (1993)
Colorants: fabric dyes on cotton and on silk					
Brazilin, Chinese tannin, Purpurin, Myricetin	CM	PO 12 months, indoor environment, on cotton and on silk	X–4.6 (up to 110)	2	Kadokura et al. (1988)
Colorants: plant fabric dyes on cotton and on silk					
Akane (pseudopurpurin)	CM	65% RH, PO 10 days, mordant Al, on silk and on cotton	19 000–X	X	Saito et al. (1994)
	CM	90% RH, PO 10 days, mordant Al, on cotton	X–1900	30	Saito et al. (1994)
	CM	90% RH, PO 10 days, mordant Al, on silk	X–1900	10	Saito et al. (1994)
Benibana	CM	65% RH, PO 10 days, mordant Al, on silk	1900–X	X	Saito et al. (1994)
	CM	90% RH, PO 10 days, mordant Al, on silk	X–1900	5	Saito et al. (1994)
Carminic acid	CM	Indoor environment, PO 10 days, on cotton	X–70	4	Saito et al. (1993)
	CM	Indoor environment, PO 10 days, on silk	X–70	50	Saito et al. (1993)
	CM	65% RH, PO 10 days, on cotton	X–1900	40	Saito et al. (1993)
	CM	65% RH, PO 10 days, on silk	19 000–X	X	Saito et al. (1993)
	CM	90% RH, PO 10 days, on silk	X–1900	10	Saito et al. (1993)
	CM	90% RH, PO 10 days, on cotton	X–1900	2	Saito et al. (1993)
Cochineal	CM	65% RH, PO 10 days, mordant Al, on cotton	X–1900	4	Saito et al. (1994)
	CM	65% RH, PO 10 days, mordant Al, on silk	X–1900	10	Saito et al. (1994)
	CM	90% RH, PO 10 days, mordant Al, on silk	X–1900	5	Saito et al. (1994)
	CM	90% RH, PO 10 days, mordant Al, on cotton	X–1900	2	Saito et al. (1994)
Curcumin	CM	Indoor environment, PO 10 days, on cotton	X–70	2	Saito et al. (1993)
	CM	Indoor environment, PO 10 days, on silk	X–70	4	Saito et al. (1993)
	CM	65% RH, PO 10 days, on cotton	X–1900	40	Saito et al. (1993)
	CM	65% RH, PO 10 days, on silk	X–1900	40	Saito et al. (1993)
	CM	90% RH, PO 10 days, on silk	X–1900	1	Saito et al. (1993)
	CM	90% RH, PO 10 days, on cotton	X–1900	1	Saito et al. (1993)

Airborne pollutants/objects	Methods	Exposure conditions	NOAEL– LOAEL, μg m^{-3}	LOAED, μg m^{-3} yr	References
Nitrogen dioxide (NO$_2$) (*continued*)					
Disperse blue 3 (CI 61505)	CM	Indoor environment, PO 10 days, on cotton and on silk	X–70	4	Saito et al. (1993)
Enju (rutin)	CM	65% RH, PO 10 days, mordant Al, on cotton	X–1900	1	Saito et al. (1994)
	CM	65% RH, PO 10 days, mordant Al, on silk	X–1900	10	Saito et al. (1994)
	CM	90% RH, PO 10 days, mordant Al, on cotton	X–1900	1	Saito et al. (1994)
	CM	90% RH, PO 10 days, mordant Al, on silk	X–1900	4	Saito et al. (1994)
Hematoxylin	CM	Indoor environment, PO 10 days, on cotton	X–70	1	Saito et al. (1993)
	CM	Indoor environment, PO 10 days, on silk	X–70	50	Saito et al. (1993)
	CM	65% RH, PO 10 days, on cotton	X–1900	3	Saito et al. (1993)
	CM	65% RH, PO 10 days, on silk	19 000–X	X	Saito et al. (1993)
	CM	90% RH, PO 10 days, on cotton	X–1900	1	Saito et al. (1993)
	CM	90% RH, PO 10 days, on silk	X–1900	10	Saito et al. (1993)
Indigo	CM	Indoor environment, PO 10 days, on cotton and on silk	X–70	50	Saito et al. (1993)
	CM	65% RH, PO 10 days, on silk and on cotton	19 000–X	X	Saito et al. (1993)
Logwood (hematoxylin)	CM	65% RH, PO 10 days, mordant Al, on silk	X–1900	10	Saito et al. (1994)
	CM	65% RH, PO 10 days, mordant Al, on cotton	X–1900	5	Saito et al. (1994)
	CM	90% RH, PO 10 days, mordant Al, on cotton	X–1900	1	Saito et al. (1994)
	CM	90% RH, PO 10 days, mordant Al, on silk	X–1900	5	Saito et al. (1994)
Quercitron	CM	Indoor environment, PO 10 days, on cotton and on silk	X–70	1	Saito et al. (1993)
	CM	65% RH, PO 10 days, on silk and on cotton	X–1900	1	Saito et al. (1993)
Suou, suo (Brazilin)	CM	65% RH, PO 10 days, mordant Al, on silk	1900–X	X	Saito et al. (1994)
	CM	65% RH, PO 10 days, mordant Al, on cotton	X–1900	40	Saito et al. (1994)
	CM	90% RH, PO 10 days, mordant Al, on cotton	X–1900	10	Saito et al. (1994)
	CM	90% RH, PO 10 days, mordant Al, on silk	X–1900	6	Saito et al. (1994)
Ukon (curcumin)	CM	65% RH, PO 10 days, mordant Al, on cotton	X–1900	10	Saito et al. (1994)
	CM	65% RH, PO 10 days, mordant Al, on silk	X–1900	30	Saito et al. (1994)
	CM	90% RH, PO 10 days, mordant Al, on silk	X–1900	3	Saito et al. (1994)
	CM	90% RH, PO 10 days, mordant Al, on cotton	X–1900	1	Saito et al. (1994)
Yamamomo (myricitrin)	CM	65% RH, PO 10 days, mordant Al, on cotton	X–1900	3	Saito et al. (1994)
	CM	65% RH, PO 10 days, mordant Al, on silk	X–1900	9	Saito et al. (1994)
	CM	90% RH, PO 10 days, mordant Al, on cotton	X–1900	3	Saito et al. (1994)
	CM	90% RH, PO 10 days, mordant Al, on silk	X–1900	9	Saito et al. (1994)
Colorants: natural organic colorants					
Aigami (malonyl awobanin)	CM	50% RH, 12 weeks, on WC paper	X–960	40	Whitmore and Cass (1989)
Bitumen	CM	50% RH, 12 weeks, on WC paper	960–X	X	Whitmore and Cass (1989)
Cochineal lake	CM	50% RH, 12 weeks, on WC paper	X–960	60	Whitmore and Cass (1989)

Airborne pollutants/objects	Methods	Exposure conditions	NOAEL–LOAEL, $\mu g\ m^{-3}$	LOAED, $\mu g\ m^{-3}$ yr	References
Nitrogen dioxide (NO$_2$) (*continued*)					
Curcumin	CM	50% RH, 12 weeks, on WC paper	X–960	20	Whitmore and Cass (1989)
Dragon's blood	CM	50% RH, 12 weeks, on WC paper	X–960	30	Whitmore and Cass (1989)
Gamboge	CM	50% RH, 12 weeks, on WC paper	X–960	50	Whitmore and Cass (1989)
Indian yellow	CM	50% RH, 12 weeks, on WC paper	960–X	X	Whitmore and Cass (1989)
Indigo	CM	50% RH, 12 weeks, on WC paper	X–960	80	Whitmore and Cass (1989)
Lac lake	CM	50% RH, 12 weeks, on WC paper	X–960	40	Whitmore and Cass (1989)
Litmus	CM	50% RH, 12 weeks, on WC paper	960–X	X	Whitmore and Cass (1989)
Weld lake	CM	50% RH, 12 weeks, on WC paper	960–X	X	Whitmore and Cass (1989)
Madder lake	CM	50% RH, 12 weeks, on WC paper	960–X	X	Whitmore and Cass (1989)
Persian berries lake	CM	50% RH, 12 weeks, on WC paper	X–960	60	Whitmore and Cass (1989)
Quercitron lake	CM	50% RH, 12 weeks, on WC paper	X–960	70	Whitmore and Cass (1989)
Saffron	CM	50% RH, 12 weeks, on WC paper	X–960	80	Whitmore and Cass (1989)
Sepia	CM	50% RH, 12 weeks, on WC paper	X–960	100	Whitmore and Cass (1989)
Van Dyke brown	CM	50% RH, 12 weeks, on WC paper	960–X	X	Whitmore and Cass (1989)
Colorants: modern synthetic organic colorants on paper, mainly watercolours					
Alizarin crimson	CM	50% RH, 12 weeks, on WC paper	X–960	80	Whitmore and Cass (1989)
Aniline black	CM	50% RH, 12 weeks, on WC paper	960–X	X	Whitmore and Cass (1989)
Arylide yellow G	CM	50% RH, 12 weeks, on WC paper	960–X	X	Whitmore and Cass (1989)
Arylide yellow 10G	CM	50% RH, 12 weeks, on WC paper	960–X	X	Whitmore and Cass (1989)
Bright red	CM	50% RH, 12 weeks, on WC paper	960–X	X	Whitmore and Cass (1989)
Dioxazine purple	CM	50% RH, 12 weeks, on WC paper	X–960	100	Whitmore and Cass (1989)
Mauve	CM	50% RH, 12 weeks, on WC paper	X–960	50	Whitmore and Cass (1989)
Naphthol	CM	50% RH, 12 weeks, on WC paper	960–X	X	Whitmore and Cass (1989)
Paliogen blue	CM	50% RH, 12 weeks, on WC paper	X–960	80	Whitmore and Cass (1989)
Paliogen yellow	CM	50% RH, 12 weeks, on WC paper	960–X	X	Whitmore and Cass (1989)
Permanent magenta	CM	50% RH, 12 weeks, on WC paper	960–X	X	Whitmore and Cass (1989)
Phthalocyanine blue	CM	50% RH, 12 weeks, on WC paper	960–X	X	Whitmore and Cass (1989)
Phthalocyanine green	CM	50% RH, 12 weeks, on WC paper	960–X	X	Whitmore and Cass (1989)
Prussian blue	CM	50% RH, 12 weeks, on WC paper	960–X	X	Whitmore and Cass (1989)
Rose carthame	CM	50% RH, 12 weeks, on WC paper	960–X	X	Whitmore and Cass (1989)
Scarlet lake	CM	50% RH, 12 weeks, on WC paper	960–X	X	Whitmore and Cass (1989)
Thioindigo violet	CM	50% RH, 12 weeks, on WC paper	960–X	X	Whitmore and Cass (1989)
Toluidine red	CM	50% RH, 12 weeks, on WC paper	960–X	X	Whitmore and Cass (1989)

Airborne pollutants/objects	Methods	Exposure conditions	NOAEL– LOAEL, $\mu g\ m^{-3}$	LOAED, $\mu g\ m^{-3}$ yr	References
Nitrogen dioxide (NO₂) *(continued)*					
Colorants: inorganic pigments					
Aureolin	CM	50% RH, 12 weeks, on WC paper	960–X	X	Whitmore and Cass (1989)
Cadmium yellow medium	CM	50% RH, 12 weeks, on WC paper	960–X	X	Whitmore and Cass (1989)
Chrome yellow	CM	50% RH, 12 weeks, on WC paper	960–X	X	Whitmore and Cass (1989)
Iron ink I	CM	50% RH, 12 weeks, on WC paper	X–960	40	Whitmore and Cass (1989)
Iron ink II	CM	50% RH, 12 weeks, on WC paper	X–960	30	Whitmore and Cass (1989)
Manganese violet	CM	50% RH, 12 weeks, on WC paper	960–X	X	Whitmore and Cass (1989)
Orpiment	CM	50% RH, 12 weeks, on WC paper	X–960	40	Whitmore and Cass (1989)
Realgar	CM	50% RH, 12 weeks, on WC paper	X–960	20	Whitmore and Cass (1989)
Vermilion	CM	50% RH, 12 weeks, on WC paper	960–X	X	Whitmore and Cass (1989)
Colorants: Japanese dyed silk					
Many colorants	CM	50% RH, 12 weeks, on WC paper	960–X	X	Whitmore and Cass (1989)
Colorants: micrograph film dyes					
Yellow	PD	80% RH, 70°C, PO 5 weeks, dye on Eastman 5272, colour negative motion-picture film	X–10 000	20	Zinn et al. (1994)
	PD	80% RH, 70°C, PO 5 weeks, dye on Eastman 52384, colour positive motion-picture film	X–10 000	90	Zinn et al. (1994)
Metals					
Copper	WG	70% RH, PO 4 weeks	X–940	30	Eriksson et al. (1993)
	WG	90% RH, PO 4 weeks	X–940	40	Eriksson et al. (1993)
Papers					
Chemical wood (acid-sized)	TS	50% RH, 1 day	X–5E6	600	Iversen and Kolar (1991)
	FE	50% RH, 1 day	X–5E6	600	Iversen and Kolar (1991)
Chemical and groundwood	FE	50% RH, 25°C, PO 4 months, 20% bleached softwood kraft pulp and 80% softwood stone groundwood	X–1900	40	Reilly et al. (2001)
	CM	50% RH, 25°C, 3 months, 20% bleached softwood kraft pulp and 80% softwood stone groundwood, colour measurement with only b value	X–1900	5	Reilly et al. (2001)
Chemical and groundwood with 5% CaCO₃	CM	50% RH, 25°C, 3 months, 20% bleached softwood kraft pulp and 80% softwood stone groundwood, colour measurement with only b value	X–1900	70	Reilly et al. (2001)
Chemical wood or cotton (neutral-sized)	TS	50% RH, 1 day	5E6–X	X	Iversen and Kolar (1991)
	FE	50% RH, 1 day	5E6–X	X	Iversen and Kolar (1991)
Cotton rag	CM	50% RH, 25°C, 60 days, colour measurement with only b value	38 000–X	X	Reilly et al. (2001)

Airborne pollutants/objects	Methods	Exposure conditions	NOAEL–LOAEL, $\mu g\ m^{-3}$	LOAED, $\mu g\ m^{-3}$ yr	References
Nitrogen dioxide (NO₂) (*continued*)					
Cotton rag	FE	50% RH, 25°C, PO 60 days	X–1900	30	Reilly et al. (2001)
Cotton rag with 5% CaCO₃	CM	50% RH, 25°C, 60 days, colour measurement with only b value	X–38 000	7000	Reilly et al. (2001)
Wood-containing paper (acid-sized)	TS	50% RH, 1 day	5E6–X	X	Iversen and Kolar (1991)
	FE	50% RH, 1 day	5E6–X	X	Iversen and Kolar (1991)
Silver image	SD	50% RH, PO 20 h, on triacetate film	X–1.9E6	2000	Carroll and Calhoun (1955)
Nitrous oxide (N₂O)					
Silver image	SD	100% RH, PO 7 days, on triacetate support	48E6–X	X	Carroll and Calhoun (1955)
Oxygen (O₂)					
Colorants	RM	Photo-oxidation of samples exposed to 10 000 lx from fluorescent tubes without UV filter	X–2.8E8	X	Arney et al. (1979)
	CM	Samples exposed to 10 000 lx from fluorescent tubes without UV filter, PO 12 000 h	X–2.8E8	X	Saunders and Kirby (1994)
Ozone (O₃)					
Colorants: top 6 most sensitive					
Curcumin	CM	50% RH, PO 12 weeks, WC on paper	X–150	0.1	Whitmore et al. (1987)
Jiang huang	CM	50% RH, PO 22 weeks, on WC paper	X–800	1	Lynn et al. (2000)
Blue 324	CM	65±2% RH, PO 12 days, on nylon 6,6	X–400	1	Ladisch et al. (1997)
Zi cao	CM	50% RH, PO 22 weeks, on WC paper	X–800	2	Lynn et al. (2000)
Alizarin crimson	CM	52% RH, 12 weeks, acrylic background binder only	X–80	2	Cass et al. (1988)
Disperse blue 3	CM	90% RH, 40°C, PO 3 days, CIDB-3 on nylon 6	X–400	0.07	Moore et al. (1984)
Colorants					
Alizarin	CM	49.5% RH, 95 days, on WC paper	X–710	20	Grosjean et al. (1987)
Alizarin carmine	RM	46±6% RH, 90 days, on WC paper	X–620	100	Drisko et al. (1985/86)
Alizarin crimson	RM	47±8% RH, 90 days, on WC paper	X–620	200	Shaver and Cass (1983)
Alizarin crimson	CM	49.5% RH, 95 days, dry powder on WC paper	X–710	8	Grosjean et al. (1987)
Alizarin crimson	CM	49.5% RH, 95 days, from tube WC on WC paper	X–710	30	Grosjean et al. (1987)
Bitumen	CM	50% RH, PO 12 weeks, WC on paper	X–150	2	Whitmore et al. (1987)
Brown madder	RM	46±6% RH, 90 days, on WC paper	X–620	100	Drisko et al. (1985/86)
Carmine	RM	46±6% RH, 90 days, on WC paper	X–620	100	Drisko et al. (1985/86)
Cochineal lake	CM	50% RH, PO 12 weeks, WC on paper	X–150	1	Whitmore et al. (1987)

Airborne pollutants/objects	Methods	Exposure conditions	NOAEL– LOAEL, $\mu g\ m^{-3}$	LOAED, $\mu g\ m^{-3}$ yr	References
Ozone (O₃) (*continued*)					
Crimson lake	RM	46±6% RH, 90 days, on WC paper	X–620	60	Drisko et al. (1985/86)
Crimson lake	RM	47±8% RH, 90 days, on WC paper	X–620	100	Shaver and Cass (1983)
Curcumin	CM	50% RH, PO 12 weeks, WC on paper	X–150	0.1	Whitmore et al. (1987)
Dragon's blood	CM	50% RH, PO 12 weeks, WC on paper	X–150	0.4	Whitmore et al. (1987)
Gamboge	CM	50% RH, PO 12 weeks, WC on paper	X–150	1	Whitmore et al. (1987)
Mauve	RM	46±6% RH, 90 days, on WC paper	X–620	100	Drisko et al. (1985/86)
Mauve	RM	47±8% RH, 90 days, on WC paper	X–620	100	Shaver and Cass (1983)
Indian yellow	CM	50% RH, PO 12 weeks, WC on paper	X–790	30	Whitmore et al. (1987)
Indigo	RM	46±6% RH, 90 days, on WC paper	X–620	100	Drisko et al. (1985/86)
Indigo	CM	50% RH, PO 12 weeks, WC on paper	X–150	0.6	Whitmore et al. (1987)
Lac lake	CM	50% RH, PO 12 weeks, WC on paper	X–150	1	Whitmore et al. (1987)
Litmus	CM	50% RH, PO 12 weeks, WC on paper	X–150	2	Whitmore et al. (1987)
Madder lake	CM	50% RH, PO 12 weeks, WC on paper	X–150	1	Whitmore et al. (1987)
Payne's grey	RM	46±6% RH, 90 days, on WC paper	X–620	100	Drisko et al. (1985/86)
Permanent rose	RM	46±6% RH, 90 days, on WC paper	620–X	X	Drisko et al. (1985/86)
Persian berries lake	CM	50% RH, PO 12 weeks, WC on paper	X–150	1	Whitmore et al. (1987)
Purple lake	RM	46±6% RH, 90 days, on WC paper	X–620	100	Drisko et al. (1985/86)
Purple lake	RM	47±8% RH, 90 days, on WC paper	X–620	200	Shaver and Cass (1983)
Quercitron lake	CM	50% RH, PO 12 weeks, WC on paper	X–150	4	Whitmore et al. (1987)
Saffron	CM	50% RH, PO 12 weeks, WC on paper	X–150	2	Whitmore et al. (1987)
Sepia	CM	50% RH, PO 12 weeks, WC on paper	150–X	X	Whitmore et al. (1987)
Van Dyke brown	CM	50% RH, PO 12 weeks, WC on paper	150–X	X	Whitmore et al. (1987)
Weld lake	CM	50% RH, PO 12 weeks, WC on paper	X–150	2	Whitmore et al. (1987)
Windsor green	RM	46±6% RH, 90 days, on WC paper	620–X	X	Drisko et al. (1985/86)
Windsor yellow	RM	46±6% RH, 90 days, on WC paper	X–620	200	Drisko et al. (1985/86)
Colorants: Chinese plant dyes					
Ban lan gen (indigo yellow)	CM	50% RH, PO 22 weeks, on silk	X–800	80	Lynn et al. (2000)
Ban lan gen	CM	50% RH, PO 22 weeks, on WC paper	X–800	50	Lynn et al. (2000)
Da qing ye (indigo leaf)	CM	50% RH, PO 22 weeks, on silk	X–800	200	Lynn et al. (2000)
Hong cha ye (black tea)	CM	50% RH, PO 22 weeks, on silk	X–800	200	Lynn et al. (2000)
Hong cha ye	CM	50% RH, PO 22 weeks, on WC paper	X–800	200	Lynn et al. (2000)
Huang bai (chinese yellow cork tree)	CM	50% RH, PO 22 weeks, on silk	X–800	300	Lynn et al. (2000)
Huang bai (gardenia)	CM	50% RH, PO 22 weeks, on WC paper	X–800	50	Lynn et al. (2000)
Huang zhi zi	CM	50% RH, PO 22 weeks, on silk	X–800	20	Lynn et al. (2000)
Jiang huang (turmeric)	CM	50% RH, PO 22 weeks, on silk	X–800	100	Lynn et al. (2000)

Airborne pollutants/objects	Methods	Exposure conditions	NOAEL– LOAEL, $\mu g\, m^{-3}$	LOAED, $\mu g\, m^{-3}$ yr	References
Ozone (O$_3$) (*continued*)					
Jiang huang	CM	50% RH, PO 22 weeks, on WC paper	X–800	1	Lynn et al. (2000)
Jiang xiang (dalbergia wood)	CM	50% RH, PO 22 weeks, on silk	X–800	80	Lynn et al. (2000)
Jiang xiang	CM	50% RH, PO 22 weeks, on WC paper	X–800	50	Lynn et al. (2000)
Ju zi pi (tangerine peel)	CM	50% RH, PO 22 weeks, on silk	800–X	X	Lynn et al. (2000)
Ju zi pi	CM	50% RH, PO 22 weeks, on WC paper	X–800	5	Lynn et al. (2000)
Su mu (sappan wood)	CM	50% RH, PO 22 weeks, on silk	X–800	90	Lynn et al. (2000)
Su mu	CM	50% RH, PO 22 weeks, on WC paper	X–800	30	Lynn et al. (2000)
Wu bei zi (Chinese gall)	CM	50% RH, PO 22 weeks, on silk	X–800	300	Lynn et al. (2000)
Wu bei zi	CM	50% RH, PO 22 weeks, on WC paper	X–800	8	Lynn et al. (2000)
Zi cao (gromwell)	CM	50% RH, PO 22 weeks, on silk	X–800	8	Lynn et al. (2000)
Zi cao	CM	50% RH, PO 22 weeks, on WC paper	X–800	2	Lynn et al. (2000)
Zi ding cao (violet)	CM	50% RH, PO 22 weeks, on silk	X–800	20	Lynn et al. (2000)
Zi ding cao	CM	50% RH, PO 22 weeks, on WC paper	X–800	4	Lynn et al. (2000)
Colorants: traditional Japanese colorants					
Many colorants	CM	50% RH, PO 12 weeks, on WC papers, on silk and on Japanese woodblock print	X–800	Various	Whitmore and Cass (1988)
Colorants: acid dyes					
Blue 40	CM	65±2% RH, PO 12 days, on nylon 6,6	X–400	4	Ladisch and Rau (1997)
Blue 127	CM	65±2% RH, PO 12 days, on nylon 6,6	X–400	2	Ladisch and Rau (1997)
Blue 232	CM	65±2% RH, PO 12 days, on nylon 6,6	X–400	2	Ladisch and Rau (1997)
Blue 324	CM	65±2% RH, PO 12 days, on nylon 6,6	X–400	1	Ladisch and Rau (1997)
Yellow 17	CM	65±2% RH, PO 12 days, on nylon 6,6	X–400	4	Ladisch and Rau (1997)
Micrograph film dyes					
Magenta	PD	50% RH, 30°C, PO 5 weeks, dye on Ilfochrome colour micrograph film CMM	10 000–X	X	Zinn et al. (1994)
Yellow	PD	50% RH, 30°C, PO 5 weeks, dye on Eastman 5272, colour negative motion-picture film	X–10 000	300	Zinn et al. (1994)
	PD	50% RH, 30°C, PO 5 weeks, dye on Eastman 52384, colour positive motion-picture film	X–10 000	400	Zinn et al. (1994)
	PD	50% RH, 30°C, PO 5 weeks, dye on Kodak Ektatherm dye diffusion print material (no gelatine)	X–10 000	30	Zinn et al. (1994)
Silver image	PD	50% RH, 30°C, PO 5 weeks, black-and-white micrograph films	10 000–X	X	Zinn et al. (1994)
	VO	50% RH, 30°C, PO 5 weeks, black-and-white micrograph films with unknown amount of H$_2$S	X–10 000	X	Zinn et al. (1994)
	VO	50% RH, indoor environment, PO 1 yr, microfilm samples	X–50	X	Zinn et al. (1994)

Airborne pollutants/objects	Methods	Exposure conditions	NOAEL–LOAEL, $\mu g\ m^{-3}$	LOAED, $\mu g\ m^{-3}$ yr	References
Ozone (O_3) (*continued*)					
Alizarin crimson	CM	52% RH, 12 weeks, acrylic background binder only	X–80	2	Cass et al. (1988)
	CM	52% RH, 12 weeks, pigmented acrylic binder applied on acrylic background binder	X–80	8	Cass et al. (1988)
	CM	52% RH, 12 weeks, acrylic film applied on dry powder with an acrylic background binder	X–80	50	Cass et al. (1988)
	CM	52% RH, 12 weeks, acrylic film applied on pigmented acrylic binder with an acrylic background binder	X–80	60	Cass et al. (1988)
Colorants: dyes on nylon					
Disperse blue 3	CM	90% RH, 40°C, PO 3 days, CIDB-3 on nylon 6	X–400	0.07	Moore et al. (1984)
	CM	90% RH, 40°C, PO 3 days, CIDB-3 on nylon 6,6	X–400	0.5	Moore et al. (1984)
Cotton	TS	X% RH, PO 50 days	80–X	X	Jaffe (1967)
	TS	X% RH, PO 50 days, wet cotton	X–80	3	Jaffe (1967)
Paper					
Chemical and groundwood	FE	50% RH, 25°C, 4 months, 20% bleached softwood kraft pulp and 80% softwood stone groundwood	X–2000	100	Reilly et al. (2001)
Chemical and groundwood with $CaCO_3$	CM	50% RH, 25°C, 4 months, 20% bleached softwood kraft pulp and 80% softwood stone groundwood, colour measurement with only b value	2000–X	X	Reilly et al. (2001)
Polymers					
Vulcanized natural rubber	VO	X% RH, PO few hours, exposure under stresses (without antioxidant)	X–40	0.005	Jaffe (1967)
	VO	X% RH, PO few hours, exposure without stresses (without antioxidant)	X–X	X	Jaffe (1967)
	VO	X% RH, 16 h, exposure without stresses (without antioxidant)	3E6–X	X	EC and HC (1999)
Polyisoprene	VO	X% RH, 2 h, exposure under stresses	X–36 000	0.8	EC and HC (1999)
Ozone, nitrogen dioxide, and PAN, see also Ozone and Nitrogen dioxide					
Colorants					
Acid red 37	CM	46% RH, 8 weeks, on Whatman paper, NO_2, PAN, O_3	X–100, 60, 80	10, 10 ,10	Grosjean et al. (1993)
Basic fushin	CM	46% RH, 8 weeks, on Whatman paper, NO_2, PAN, O_3	X–100, 60, 80	3, 2, 3	Grosjean et al. (1993)
Brilliant green	CM	46% RH, on Whatman paper, NO_2, PAN, O_3, PO 8 weeks	X–100, 60, 80	3, 2, 3	Grosjean et al. (1993)
Curcumin	CM	46% RH, on WCs papers, NO_2, PAN, O_3, PO 8 weeks	X–100, 60, 80	0.4, 0.2, 0.3	Grosjean et al. (1993)

Airborne pollutants/objects	Methods	Exposure conditions	NOAEL–LOAEL, $\mu g\ m^{-3}$	LOAED, $\mu g\ m^{-3}$ yr	References
Ozone, nitrogen dioxide, and PAN, see also Ozone and Nitrogen dioxide (*continued*)					
Disperse blue 3	CM	46% RH, 8 weeks, on WC papers, NO_2, PAN, O_3	X–100, 60, 80	3, 2, 3	Grosjean et al. (1993)
Indigo	CM	46% RH, on WC papers, NO_2, PAN, O_3, PO 8 weeks	X–100, 60, 80	0.2, 0.1, 0.1	Grosjean et al. (1993)
Indigo carmine	CM	46% RH, 8 weeks, on Whatman paper, NO_2, PAN, O_3	X–100, 60, 80	20, 10, 20	Grosjean et al. (1993)
Pararosaniline base	CM	46% RH, 8 weeks, on Whatman paper, NO_2, PAN, O_3	X–100, 60, 80	3, 2, 3	Grosjean et al. (1993)
Reactive blue 2	CM	46% RH, on Whatman paper, NO_2, PAN, O_3, PO 8 weeks	X–100, 60, 80	3, 2, 3	Grosjean et al. (1993)
For less fugitive colorants; see reference					Grosjean et al. (1993)
Particles (fine; $PM_{2.5}$)					
Horizontal surfaces	VO, M	Indoor environment, floors, based on elemental carbon deposition	X–(4.1–50)	10	Bellan et al. (2000); Nazaroff et al. (1993)
Vertical surfaces	VO, M	Indoor environment, walls, based on elemental carbon deposition	X–(4.1–50)	50	Bellan et al. (2000); Nazaroff et al. (1993)
Peroxyacetyl nitrate, PAN ($CH_3C(O)OONO_2$)					
Colorants					
Curcumin	CM	45% RH, 12 weeks, WC papers	X–140	20	Williams et al. (1993a)
Basic fuchsin	CM	45% RH, 12 weeks, WC papers	X–140	10	Williams et al. (1993a)
Twelve colorants	CM	45% RH, 12 weeks, WC papers	140–X	X	Williams et al. (1993a)
Sulphur dioxide (SO_2)					
Calcium carbonate-based materials					
Coral	VO	Conditions unknown; reported from the Canadian Museum of Nature	X–X	X	Waller (2002)
Limestone	WG	90% RH, PO 6 weeks	X–4300	90	Johansson et al. (1988)
	GYC	100% RH, PO 20 days	2700–X	X	Moroni and Poli (1996)
	GYC	100% RH, PO 20 days	X–13 000	80	Moroni and Poli (1996)
	WL	X% RH, outdoor conditions: coastal site, 200 days	X–11	20	Cooke and Gibbs (1994)
	WL	X% RH, outdoor conditions: inland site, 200 days	X–21	30	Cooke and Gibbs (1994)
Marble	FT/SEM, M	100% RH, PO 188 days, LD: 20 μm	X–27 000	70	Lal Gauri et al. (1989)
Marble	WG	90% RH, PO 6 weeks	X–4300	200	Johansson et al. (1988)
Travertine	WG	90% RH, PO 6 weeks	X–4300	200	Johansson et al. (1988)

Airborne pollutants/objects	Methods	Exposure conditions	NOAEL–LOAEL, $\mu g\ m^{-3}$	LOAED, $\mu g\ m^{-3}$ yr	References

Sulphur dioxide (SO_2) (*continued*)

Colorants

Curcumin	CM	46% RH, 12 weeks, on WC paper	X–140	10	Williams et al. (1993b)
Indigo	CM	46% RH, 12 weeks, on Whatman paper	X–140	20	Williams et al. (1993b)
Mauve	CM	46% RH, 12 weeks, on WC paper	X–140	10	Williams et al. (1993b)
Reactive blue 2 (Procion blue HB)	CM	46% RH, 12 weeks, on Whatman paper	X–140	20	Williams et al. (1993b)
Basic fuchsin	CM	46% RH, PO 12 weeks, on Whatman paper	X–140	10	Williams et al. (1993b)
	CM	50% RH, PO 67 days, on WC paper	X–450	10	Tétreault and Lai (2001)
Brilliant green	CM	46% RH, PO 12 weeks, on Whatman paper	X–140	10	Williams et al. (1993b)
Parasoaniline base	CM	46% RH, PO 12 weeks, on Whatman paper	X–140	10	Williams et al. (1993b)
Other colorants	CM	46% RH, 12 weeks, on WC paper	140–X	X	Williams et al. (1993b)

Leathers, vegetable-tanned

Leathers, vegetable-tanned	SC	Indoor environment, 60 yrs, in countryside environment, EL	X–0.6	10	Larsen (1997); Brimblecombe (1997); Blades (1997)
	VO	Indoor environment, 60 yrs, in countryside environment, EL	X–0.6	40	Brimblecombe (1997); Blades (1997)
	SC	Indoor environment, 60 yrs, compared with countryside environment, EL	X–17	70	Larsen (1997); Brimblecombe (1997); Blades (1997)
	SC	Indoor environment, PO 7 yrs, calf leather, EL	X–60	40	Chahine (1991)
	SC	90% RH, PO 12 weeks, calf leather	X–65 000	900	Chahine (1991)

Metals

Copper	AES	Indoor environment, PO 10 yrs (outdoor avg. level: 53, I/O = 0.88)	X–47	500	Graedel (1987b); Schubert and D'Egidio (1990); Hisham et al. (1991b)
	WG	70% RH, PO 4 weeks	X–1300	70	Eriksson et al. (1993)
	WG	90% RH, PO 4 weeks	X–1300	9	Eriksson et al. (1993)

Micrograph film dyes

Azo-naphthol (cyan)	VO	72 h	X–11 000	90	Leyshon and Holstead (1988)
Magenta	PD	50% RH, 30°C, PO 5 weeks, dye on Ilfochrome colour micrograph film CMM	13 000–X	X	Zinn et al. (1994)
Yellow	PD	50% RH, 30°C, PO 5 weeks, dye on Eastman 5272, colour negative motion-picture film	13 000–X	X	Zinn et al. (1994)
	PD	50% RH, 30°C, PO 5 weeks, dye on Eastman 52384, colour positive motion-picture film	13 000–X	X	Zinn et al. (1994)
	PD	50% RH, 30°C, PO 5 weeks, dye on Kodak Ektatherm dye diffusion print material (no gelatine)	13 000–X	X	Zinn et al. (1994)

Papers

Chemical wood	RTT	58% RH, PO 180 days, contains also additives	X–27 000	800	Edwards et al. (1968)
	RTT	58% RH, PO 180 days, papers made of 80% groundwood, 20% chemical wood	X–27 000	200	Edwards et al. (1968)

Airborne pollutants/objects	Methods	Exposure conditions	NOAEL–LOAEL, μg m^{-3}	LOAED, μg m^{-3} yr	References
Sulphur dioxide (SO$_2$) (*continued*)					
Chemical and groundwood with CaCO$_3$	CM	50% RH, 25°C, 4 months, 20% bleached softwood kraft pulp and 80% softwood stone groundwood, colour measurement with only b value	1300–X	X	Reilly et al. (2001)
Cotton rag	pH	Indoor environment, avg. 190–230 yrs	X–47	2000–6000	Lyth Hudson (1967)
Cotton rag (purest)	RTT	58% RH, PO 180 days	X–27 000	20 000	Edwards et al. (1968)
Cotton rag	RTT	58% RH, PO 180 days, contains also additives	X–27 000	800	Edwards et al. (1968)
Newsprint	SC	62% RH, PO 29 weeks (100% chemical wood, colophane, starch, kaolin)	X–140	30	Williams and Grosjean (1990)
Silver image	PD	50% RH, 30°C, PO 5 weeks, black-and-white micrograph films	13 000–X	X	Zinn et al. (1994)
Sulphur dioxide (SO$_2$) and nitrogen dioxide (NO$_2$)					
Calcium carbonate-based materials					
Limestone	GYC	100% RH, PO 20 days	SO$_2$: 2700–X NO$_2$: 190–X	X X	Moroni and Poli (1996)
	WG	90% RH, PO 6 weeks	SO$_2$: X–4300 NO$_2$: X–5900	30 40	Johansson et al. (1988)
	GYC	100% RH, PO 20 days	SO$_2$: X–13 000 NO$_2$: X–380	60 2	Moroni and Poli (1996)
Marble	WG	90% RH, PO 6 weeks	SO$_2$: X–4300 NO$_2$: X–5900	40 50	Johansson et al. (1988)
Travertine	WG	90% RH, PO 6 weeks	SO$_2$: X–4300 NO$_2$: X–5900	30 40	Johansson et al. (1988)
Gelatine on photography	CHR	50%RH, PO 30 days, low tanned gelatine	SO$_2$: X–27E3 NO$_2$: X–38E3	300–500 400–700	Nguyen et al. (1999)
Leathers, vegetable-tanned	pH	PO 7 yrs, indoor environment, calf leather, EL	SO$_2$: X–60 NO$_2$: X–18	30 9	Chahine (1991)
	pH	90% RH, PO 12 weeks, calf leather	SO$_2$: X–65E3 NO$_2$: X–18E3	1000 300	Chahine (1991)
Metals					
Copper	WG	70% RH, PO 4 weeks	SO$_2$: X–1300 NO$_2$: X–940	30 20	Eriksson et al. (1993)
	WG	90% RH, PO 4 weeks	SO$_2$:X–1300 NO$_2$: X–940	2 2	Eriksson et al. (1993)
Papers					
Bleach kraft	DP	50% RH, 4 days, aspen or spruce	SO$_2$: X–27E3 NO$_2$: X–38E3	10 20	Bégin et al. (1999)
	FE	50% RH, 4 days, aspen or spruce	SO$_2$: X–27E3 NO$_2$: X–38E3	20 20	Bégin et al. (1999)

Airborne pollutants/objects	Methods	Exposure conditions	NOAEL–LOAEL, $\mu g\ m^{-3}$	LOAED, $\mu g\ m^{-3}$ yr	References
		Sulphur dioxide (SO₂) and nitrogen dioxide (NO₂) (*continued*)			
	DP	50% RH, 4 days, softwood	SO_2: X–27E3 NO_2: X–38E3	40 30	Havermans (1997)
	TS	50% RH, 4 days, aspen or spruce	SO_2: X–27E3 NO_2: X–38E3	60 20	Bégin et al. (1999)
	pH	50% RH, 4 days, aspen or spruce	SO_2: X–27E3 NO_2: X–38E3	30 50	Bégin et al. (1999)
	SC	50% RH, 4 days, aspen or spruce	SO_2: X–27E3 NO_2: X–38E3	100 100	Bégin et al. (1999)
	SC	50% RH, 4 days, softwood	SO_2: X–27E3 NO_2: X–38E3	100 200	Havermans (1997)
	CI/IS	50% RH, 4 days, softwood	SO_2: X–27E3 NO_2: X–38E3	200 400	Havermans (1997)
	pH	50% RH, 28°C, PO 9 weeks (100% chemical wood, colophane, starch, kaolin)	SO_2: X–67E3 NO_2: X–19E3	300 70	Daniel et al. (1988)
	FE	50% RH, 28°C, PO 9 weeks (100% chemical wood, colophane, starch, kaolin)	SO_2: X–67E3 NO_2: X–19E3	600 100	Daniel et al. (1988)
	FE	50% RH, 28°C, PO 9 weeks (100% chemical wood, colophane, starch, kaolin)	SO_2: X–67E3 NO_2: X–19E3	1000 300	Daniel et al. (1988)
	pH	90% RH, 28°C, PO 9 weeks, preacidified (100% bleached chemical wood, colophane, starch, kaolin)	SO_2: X–67E3 NO_2: X–19E3	1000 300	Daniel et al. (1991)
	DP	90% RH, 28°C, PO 9 weeks (100% bleached chemical wood, colophane, starch, kaolin)	SO_2: X–67E3 NO_2: X–19E3	3000 600	Daniel et al. (1991)
	SC	50% RH, 28°C, PO 9 weeks (100% chemical wood, colophane, starch, kaolin)	SO_2: X–67E3 NO_2: X–19E3	4000 1000	Daniel et al. (1988)
Buffered bleach kraft	SC	50% RH, 4 days, aspen or spruce, 2% $CaCO_3$	SO_2: X–27E3 NO_2: X–38E3	50 80	Bégin et al. (1999)
	DP	50% RH, 4 days, aspen or spruce, 2% $CaCO_3$	SO_2: X–27E3 NO_2: X–38E3	100 100	Bégin et al. (1999)
	TS	50% RH, 4 days, aspen or spruce, 2% $CaCO_3$	SO_2: X–27E3 NO_2: X–38E3	100 200	Bégin et al. (1999)
	pH	50% RH, 4 days, aspen or spruce, 2% $CaCO_3$	SO_2: X–27E3 NO_2: X–38E3	1000 1000	Bégin et al. (1999)
	FE	50% RH, 4 days, aspen or spruce, 2% $CaCO_3$	SO_2: 27E3–X NO_2: 38E3–X	X X	Bégin et al. (1999)
Chemical and groundwood	FE	50% RH, 25°C, PO 6 months, 20% bleached softwood kraft pulp and 80% softwood stone groundwood	SO_2: X–130 NO_2: X–190	60 90	Reilly et al. (2001)
	CM	50% RH, 25°C, PO 6 months, 20% bleached softwood kraft pulp and 80% softwood stone groundwood, colour measurement with only b value	SO_2: X–130 NO_2: X–190	10 10	Reilly et al. (2001)

Airborne pollutants/objects	Methods	Exposure conditions	NOAEL–LOAEL, $\mu g\ m^{-3}$	LOAED, $\mu g\ m^{-3}\ yr$	References
Sulphur dioxide (SO₂) and nitrogen dioxide (NO₂) *(continued)*					
Cotton rag	DP	50% RH, 4 days	SO₂: X–27E3 NO₂: X–38E3	10 10	Bégin et al. (1999)
	FE	50% RH, 4 days	SO₂: X–27E3 NO₂: X–38E3	10 20	Bégin et al. (1999)
	DP	50% RH, 4 days, 95% cotton and 5% softwood	SO₂: X–27E3 NO₂: X–38E3	20 30	Havermans (1997)
	SC	50% RH, 4 days, 95% cotton and 5% softwood	SO₂: X–27E3 NO₂: X–38E3	20 30	Havermans (1997)
	TS	50% RH, 4 days	SO₂: X–27E3 NO₂: X–38E3	70 30	Bégin et al. (1999)
	SC	50% RH, 4 days	SO₂: X–27E3 NO₂: X–38E3	30 40	Bégin et al. (1999)
	pH	50% RH, 4 days	SO₂: X–27E3 NO₂: X–38E3	30 50	Bégin et al. (1999)
	DP	90% RH, 28°C, PO 9 weeks, Whatman paper, preacidified	SO₂: X–67E3 NO₂: X–19E3	100 30	Daniel et al. (1991)
	CI/IS	50% RH, 4 days, 95% cotton and 5% softwood	SO₂: X–27E3 NO₂: X–38E3	300 400	Havermans (1998)
	pH	90% RH, 28°C, PO 9 weeks, Whatman paper, preacidified	SO₂: X–67E3 NO₂: X–19E3	2000 400	Daniel et al. (1991)
Cotton rag with 5% CaCO₃	CM	50% RH, 25°C, PO 6 months, colour measurement with only b value	SO₂: 130–X NO₂: 190–X	X X	Reilly et al. (2001)
Newsprint	FE	50% RH, 28°C, PO 9 weeks, 60% mechanic pulp and 40% bleached chemical wood	SO₂: X–67E3 NO₂: X–19E3	100 30	Daniel et al. (1988)
	pH	90% RH, 28°C, PO 9 weeks, preacidified, 60% mechanic pulp and 40% bleached chemical wood	SO₂: X–67E3 NO₂: X–19E3	100 30	Daniel et al. (1991)
	pH	50% RH, 28°C, PO 9 weeks, 60% mechanic pulp and 40% bleached chemical wood	SO₂: X–67E3 NO₂: X–19E3	200 40	Daniel et al. (1988)
	SC	50% RH, 28°C, PO 9 weeks, 60% mechanic pulp and 40% bleached chemical wood	SO₂: X–67E3 NO₂: X–19E3	500 10	Daniel et al. (1988)
	SC	90% RH, 28°C, PO 9 weeks, preacidified, 60% mechanic pulp and 40% bleached chemical wood	SO₂: X–67E3 NO₂: X–19E3	500 100	Daniel et al. (1991)
	CM	50% RH, 28°C, PO 9 weeks, 60% mechanic pulp and 40% bleached chemical wood	SO₂: X–67E3 NO₂: X–19E3	600 100	Daniel et al. (1988)

			%RH	%RH yr	
Water vapour (H₂O)					
Cellulose triacetate film	pH	Accelerated ageing: 6 yrs, end point at pH 4.0	X–50	300	Adelstein et al. (1995)
	Break strain	Accelerated ageing: 70 yrs, end point at 66% retention	X–50	3500	Adelstein et al. (1995)
	pH	Accelerated ageing, criteria: free acidity level of 0.5	X–20	2000	Reilly (1993)

Airborne pollutants/objects	Methods	Exposure conditions	NOAEL– LOAEL, μg m^{-3}	LOAED, μg m^{-3} yr	References
Water vapour (H₂O) (*continued*)			%RH	%RH yr	
Dyes of colour slide, negatives and prints	CM	Accelerated ageing, criteria: 30% loss of the most sensitive dyes	X–20	2000	Reilly (1998)
Glass (rich soda content)	VO	Indoor environment	X–X	X	Riederer (1997); Oakley (1990); Ryan et al. (1993)
Organic material	VO	Indoor environment, no mould growth after more than 10 yrs	60–X	X	Michalski 2000
Magnetic tape	VO	Indoor environment (sticky tape syndrome); also dry tape syndrome	X–50	X	Howarth (1998)
	Usable	Indoor environment, 12–14 yrs	X–50	700	Howarth (1998)
Poly(ethylene terephthalate) film	Usable	Accelerated ageing: 500 yrs	X–50	25 000	Smith (1991)
	Stress	Emulsion-coated film, accelerated aging: 2000 yrs, end point at 66% retention	X–50	100 000	Adelstein et al. (1995)
Videotape (urethane)	Usable	Indoor environment, 40 yrs	X–50	2000	Rodgers (1998)
	DH	Accelerated ageing: 30 yrs	X–50	1500	Van Bogart (1995)
	Peel force	Accelerated ageing: 20 yrs	X–50	1000	Smith (1991)

Appendix 3. Effects of Mixtures of Pollutants on Deterioration[a]

Pollutants ($\mu g\ m^{-3}$)	Materials	Effect on LOAED[b] ($\mu g\ m^{-3}$ yr)	Methods and conditions	References
Effect of different gases on the deterioration of copper and silver by hydrogen sulphide				
28 H_2S + 190 NO_2	Copper	0.9 (H_2S) to 0.5 (H_2S + NO_2)	EA, 75% RH, RT[c], 10 days	Fiaud and Guinement (1986)
140 H_2S + 290 Cl_2	Copper	0.9 to 1.0 (Cl_2 alone: 0.6)	EA, 75% RH, RT, 10 days	Fiaud and Guinement (1986)
3300 H_2S + 400 O_3	Copper	2 to 1	EDXA, 93% RH, RT, 2 days	Franey (1988)
3300 H_2S + 400 O_3 + sunlight	Copper	2 to 1	EDXA, 93% RH, RT, 2 days	Franey (1988)
3300 H_2S + 340 O_3	Copper	2 to 1	EDXA, 93% RH, RT, 2 days	Gradel et al. (1984)
3300 H_2S + light	Copper	2 to 1	EDXA, 93% RH, RT, 2 days	Gradel et al. (1984)
3300 H_2S + 400 O_3 + light	Copper	2 to 1	EDXA, 93% RH, RT, 2 days	Gradel et al. (1984)
140 H_2S + 3800 NO_2	Silver	2 to 2	EA, 75% RH, RT, 10 days	Fiaud and Guinement (1986)
140 H_2S + 290 Cl_2	Silver	2 to 1	EA, 75% RH, RT, 10 days	Fiaud and Guinement (1986)
28 H_2S + 5300 SO_2	Silver	0.01 to 0.01	WG, dry and 75% RH, 21 days	Pope et al. (1968)
1400 H_2S + 2700 SO_2	Silver	0.7 to 0.5	CR, 85% RH, 30°C, 2.5 days	Lorenzen (1971)
1400 H_2S + 2900 Cl_2	Silver	0.7 to 0.07: important shift	CR, 85% RH, 30°C, 2.5 days	Lorenzen (1971)
Effect of ozone on the deterioration of black-and-white films by hydrogen sulphide				
X H_2S + 50 O_3	Silver films	High to low: important shift	VO, 50% RH, RT, 1 yr, limited documentation	Zinn et al. (1994)
Effect of sulphur dioxide on the deterioration of papers by nitrogen dioxide				
1900 NO_2 + 1300 SO_2	Acidic paper	5 to 20 (or +): improvement	CM, 50% RH, 25°C, 4 months	Reilly et al. (2001)
1900 NO_2 + 1300 SO_2	Acidic paper	40 to 90: improvement	FE, 50% RH, 25°C, 4 months	Reilly et al. (2001)
900 NO_2 + 1300 SO_2	Cotton rag paper	30 to 300: improvement	FE, 50% RH, 25°C, 4 months	Reilly et al. (2001)
Effect of nitrogen dioxide on the deterioration of copper and limestone by sulphur dioxide				
1300 SO_2 + 940 NO_2	Copper	70 to 30	WG, 70% RH, RT, 4 weeks	Eriksson et al. (1993)
13 000 SO_2 + 380 NO_2	Limestone	80 to 60	GYC, 100% RH, 25°C, 20 days	Moroni and Poli (1996)
Effect of sea salts on the deterioration of limestone by sulphur dioxide				
SO_2 + airborne sea salt	Limestone	30 to 20	WL, X% RH, outdoor coastal versus inland sites, 200 days	Cooke and Gibbs (1994)
Effect of nitrogen dioxide and peroxyacetyl nitrate (PAN) on the deterioration of curcumin and indigo by ozone				
76 O_3 + 100 NO_2 + 60 PAN (O_3 alone: 150 $\mu g\ m^{-3}$)	Curcumin	0.1 to 0.3	CM, 45–50% RH, RT, 8 weeks, the LOAED difference is possibly lower	Grosjean et al. (1993)
76 O_3 + 100 NO_2 + 60 PAN (O_3 alone: 150 $\mu g\ m^{-3}$)	Indigo	0.6 to 0.1	CM, 45–50% RH, RT, 8 weeks, the LOAED difference is possibly lower	Grosjean et al. (1993)

Pollutants (μg m^{-3})	Materials	Effect on LOAED[b] (μg m^{-3} yr)	Methods and conditions	References
Effect of formic acid (FA) and formaldehyde (F) on the deterioration of lead by acetic acid (AA)				
500 AA + 380 FA	Lead	Improved by a factor of 5	WG, 75% RH, RT, 140 days	Tétreault et al. (2003)
27 000 AA + 15 000 FA	Lead	Improved by a factor of 10	WG, 75% RH, RT, 140 days	Tétreault et al. (2003)
27 000 AA + 15 000 FA + 3800 F	Lead	Improved by a factor of 10	WG, 75% RH, RT, 140 days	Tétreault et al. (2003)
Effect of acetic acid (AA) and formaldehyde (F) on the deterioration of copper by formic acid (FA)				
15 000 FA + 27 000 AA	Copper	No change	WG, 75% RH, RT, 140 days	Tétreault et al. (2003)
15 000 FA + 27 000 AA + 3800 F	Copper	No change	WG, 75% RH, RT, 140 days	Tétreault et al. (2003)

a: LOAED and NOAEL data, experimental conditions, and abbreviations are same as Appendix 2.
b: The first number represents the LOAED of the main pollutant.
   The second represents the new LOAED after the addition of other agents of deterioration.
c: RT: room temperature (20–25°C).

APPENDIX 4. REDUCTION OF THE LEVEL OF AIRBORNE POLLUTANTS

APPENDIX 4A. POSSIBLE REDUCTION OF AIRBORNE POLLUTANT LEVELS IN ROOMS AND ENCLOSURES

Key airborne pollutants	Reference concentration range, μg m^{-3}		Range of indoor–outdoor ratio (%) [a]		Range of inside–outside enclosure ratio (%)[d]		
	Urban area	Clean low troposphere	HVAC without filters[b] or NV[c]	HVAC with filters	N = 1/d [e]	N = 0.1/d	N = 0.1–1/d with sorbents
Acetic acid	0.5–20	0.3–5	≥100	≥100	>>100	>>100	≥ or ≤100[f]
Hydrogen sulphide	0.02–1	0.01–1	10–400	10–300	1–50	1–90	≤1
Nitrogen dioxide	3–200	0.2–20	1–90	1–30	1–50	1–90	≤1
Ozone	20–300	2–200	1–100	1–30	1–20	0.1–2	≤0.1
Sulphur dioxide	6–100	0.1–30	10–50	1–10	1–50	1–90	≤1
Fine particles; PM$_{2.5}$	1–100	1–30	10–100	10–30	1–50	0.1–5	≤1
Water vapour [g]							

a: Data based on Appendices 1 and 4B, and on Figures 34 and 38.

b: The HVAC system has no high-performance particles or gaseous filters.

c: NV = natural ventilation.

d: Based on various scenarios using deposition velocity, including possible sources of emissive products and objects.

e: An air exchange rate (N) of 1 per day is typical for a well-sealed display case. A display case with an N of 0.1 is considered a high-tech enclosure.

f: Depends if the source is inside or outside the enclosure.

g: Not covered.

APPENDIX 4B. REDUCTION OF OUTDOOR POLLUTANTS INSIDE BUILDINGS

Hydrogen sulphide (p. 131)   Ozone (p. 133)   Sulphur dioxide (p. 134)

Nitrogen dioxide (p. 132)   Particles, fine (p. 133)

Legend: NV = naturally ventilated; N = air exchange rate; STD = standard deviation (95%); LD = limit of detection; HVAC1 = HVAC system using high-performance gas and dust filters; HVAC2/NV = HVAC without special filters or using only natural ventilation (NV) (good performance may be due to low infiltration, low intake air).

Design and filter system	I/O (%)	Descriptions	References
Hydrogen sulphide		Average ranges of I/O%: HVAC1 = 10–300 and HVAC2/NV = 10–400	
Leaky room without portable carbon filter	60		Blades et al. (2000)
Leaky room with portable carbon filter	<LD		Blades et al. (2000)
Museum with HVAC and gas filters, N = 1.3/h, 80% recirculation	100	Museum of London, UK	Cassar et al. (1999)
Museum with HVAC (no special filter)	180–340	Groningen Museum, The Netherlands, urban area, summer	Ankersmit et al. (2000); Ankersmit (2001)

Design and filter system	I/O (%)	Descriptions	References
Hydrogen sulphide (*continued*)		**Average ranges of I/O%: HVAC1 = 10–300 and HVAC2/NV = 10–400**	
Museum with HVAC (no special filter)	14–64	Groningen Museum, The Netherlands, urban area, autumn	Ankersmit et al. (2000); Ankersmit (2001)
Museum with HVAC (no special filter)	22–110	Bonnefanten Museum, The Netherlands, industrial area, summer	Ankersmit et al. (2000); Ankersmit (2001)
Museum with NV	73–180	Tromp's Huys, The Netherlands, coastal area, summer	Ankersmit et al. (2000); Ankersmit (2001)
Museum with NV	200–450	Huize Doorn, The Netherlands, agricultural area, summer	Ankersmit et al. (2000); Ankersmit (2001)
Museum with NV	44–82	Huize Doorn, The Netherlands, agricultural area, autumn	Ankersmit et al. (2000); Ankersmit (2001)
Nitrogen dioxide		**Average ranges of I/O%: HVAC1 = 1–30 and HVAC2/NV = 1–90**	
Building, NV	90	Birmingham, UK, February 1996	Kukadia et al. (1996)
Large galleries, well sealed N = 2, NV	82–95	Simulated, no indoor sources	Blades et al. (2000)
Building, HVAC system, particle filter	80	Birmingham, UK, February 1996	Kukadia et al. (1996)
Large galleries, well sealed N = 1, NV	75–93	Simulated, no indoor sources	Blades et al. (2000)
Large galleries, well sealed N = 0.1/h	35–70	Simulated, no indoor sources	Blades et al. (2000)
Medium size rooms, well sealed, N = 2/h, NV	35	Simulated, no indoor sources	Blades et al. (2000)
Medium size rooms, well sealed, N = 1/h, NV	20	Simulated, no indoor sources	Blades et al. (2000)
Medium size rooms, well sealed, N = 0.1/h, NV	2	Simulated, no indoor sources	Blades et al. (2000)
Museums, HVAC system without carbon filters	50–90		Blades et al. (2000)
Museum, local filtration	10–30		Blades et al. (2000)
Museums, HVAC system with carbon filters	7–20		Blades et al. (2000)
Museums, well sealed, NV, N = 0.3/h	84	Bethnal Green Museum, UK	Cassar et al. (1999)
Museum with HVAC and gas filters, N = 1.3/h, 80% recirculation	19	Museum of London, UK	Cassar et al. (1999)
Office, NV, N = 1.6/h	90		Cassar et al. (1999)
Office, HVAC without recirculation or carbon filters, N = 1.2/h	80		Cassar et al. (1999)
Museum, NV	40–80	National Gallery of London, UK	Saunders (1993)
Museum, HVAC with gas filters	4–11	National Gallery of London, UK	Saunders (1993)
Museum, NV	50	National Gallery of London, Sainsbury Wing, UK	Saunders (1993)
Museum, HVAC with gas filters	8	National Gallery of London, Sainsbury Wing, UK	Saunders (1993)
Museum, small portable HVAC with gas filter	17–47	24-h avg., Gene Autry Museum, California, summer	Hisham and Grosjean (1991a)
Archive, storeroom, HVAC with fine dust filters	35	1-week avg., The Hague, summer	Lanting (1990)
Archive, storeroom, HVAC with fine dust filters	10	1-week avg., The Hague, winter	Lanting (1990)
Archive, storeroom, HVAC with fine dust filters, 20% recirculation	73	The Hague	Lanting (1990)
Archive, storeroom, HVAC with fine dust filters, 35% recirculation	65	The Hague	Lanting (1990)

Design and filter system	I/O (%)	Descriptions	References
Nitrogen dioxide (*continued*)	Average ranges of I/O%: HVAC1 = 1–30 and HVAC2/NV = 1–90		
Archive, storeroom, HVAC with fine dust filters, 62% recirculation	35	The Hague	Lanting (1990)
Archive, storeroom, HVAC with fine dust filters, 93% recirculation	3	The Hague	Lanting (1990)
Museum, exhibit room with tapestry, NV	32	1-week avg., Middelburg, The Netherlands, summer	Lanting (1990)
Museum, exhibit room with tapestry, NV	39	1-week avg., Middelburg, The Netherlands, winter	Lanting (1990)
Ozone	Average ranges of I/O%: HVAC1 = 1–30 and HVAC2/NV = 1–100		
Museum, high N (open doors and windows), no HVAC system	50–87	2 samples in 2 museums, 8-h avg., day, summer, California	Druzik et al. (1990)
Office building, ventilation duct, no recirculation	70–90	1 sample, 1 office, 1-h avg., at peak level, summer, California	Sabersky et al. (1973)
Office building, ventilation duct, N = about 10/h, no recirculation	50–120	1 sample, 1 office, 1-h avg., at peak level, summer, California	Sabersky et al. (1973)
Museum, no recirculation during the day	60–80	Day, summer, UK	Davies et al. (1984)
Office building, ventilation duct, 30% recirculation	55–75	Peak, summer, California	Sabersky et al. (1973)
Office building, ventilation duct, 30% recirculation	50–100	Day, summer, California	Sabersky et al. (1973)
Museum, low N, no HVAC system	10–58	Day, summer, California	Druzik et al. (1990)
Private house, natural convection, open windows and periodic open doors	60–80	Peak, summer, California	Sabersky et al. (1973)
Museum, natural convection-induced N, no HVAC system	10–58	Day, summer, California	Druzik et al. (1990)
Museum, HVAC system without gas filter	25–41	Day, summer, California	Druzik et al. (1990)
Museum, HVAC system with AC filters	4–31	Day, 8-h avg., summer, California	Druzik et al. (1990)
Museum, small HVAC with gas filter	12–42	Gene Autry Museum, California, summer	Hisham and Grosjean (1991b)
Archive, storeroom, HVAC with fine dust filters	<3	1-week avg., The Hague, summer	Lanting (1990)
Archive, storeroom, HVAC with fine dust filters, 20% recirculation	77	The Hague	Lanting (1990)
Archive, storeroom, HVAC with fine dust filters, 35% recirculation	70	The Hague	Lanting (1990)
Archive, storeroom, HVAC with fine dust filters, 62% recirculation	20	The Hague	Lanting (1990)
Archive, storeroom, HVAC with fine dust filters, 93% recirculation	3	The Hague	Lanting (1990)
Museum, exhibit room with tapestry, NV	<4	1-week avg., Middelburg, The Netherlands, summer	Lanting (1990)
Particles, fine (PM$_{2.5}$)	Average ranges of I/O%: HVAC1 = 10–30 and HVAC2/NV = 10–100		
Museum, well sealed, NV, N = 0.3/h, particles <3.5 μm	133	Bethnal Green Museum, UK	Cassar et al. (1999)
Museum, HVAC with gas filters, N = 1.3/h, 80% recirculation particles <3.5 μm	15	Museum of London, UK	Cassar et al. (1999)
Museum, NV, N = 1.6–3.6/h, particles <2.1 μm	90–100	Sepulveda House, CA, USA, summer and winter	Nazaroff et al. (1993)

Design and filter system	I/O (%)	Descriptions	References
Particles, fine (PM$_{2.5}$) (*continued*)	Average ranges of I/O%: HVAC1 = 10–30 and HVAC2/NV = 10–100		
Museum, HVAC without special dust filters, N = 0.3/h, particles <2.1 μm	71–93	Southwest Museum, based on STD, CA, USA, summer	Nazaroff et al. (1993)
Museum, HVAC without special dust filters, N = 0.3, particles <2.1 μm	34–60	Southwest Museum, based on STD, CA, USA, winter	Nazaroff et al. (1993)
Museum, HVAC with special dust filters, N = 0.4–0.7/h, particles <2.1 μm	13–26	Norton Simon Museum, based on STD, CA, USA, summer	Nazaroff et al. (1993)
Museum, HVAC with special dust filters, N = 0.4–0.7/h, particles <2.1 μm	13–24	Norton Simon Museum, based on STD, CA, USA, winter	Nazaroff et al. (1993)
Museum, HVAC without special dust filters, N = 0.3–1.0/h, particles <2.1 μm	40–80	Scott Gallery, based on STD, CA, USA, summer	Nazaroff et al. (1993)
Museum, HVAC without special dust filters, N = 0.3–1.0/h, particles <2.1 μm	13–30	Scott Gallery, based on STD, CA, USA, winter	Nazaroff et al. (1993)
Museum, HVAC without special dust filters, N = 1.2–1.3/h, particles <2.1 μm	28–54	Getty Museum, Malibu Museum based on STD, CA, USA, summer	Nazaroff et al. (1993)
Museum, HVAC without special dust filters, N = 1.2–1.3/h, particles <2.1 μm	8–94	Getty Museum, Malibu Museum, based on STD, CA, USA, winter	Nazaroff et al. (1993)
Residential building, NV, open window, particle diameter range 0.01–1 μm	40–60	Central Copenhagen, 4th floor, May–June	Wahlin et al. (2002)
Residential building, NV, without sorbing products	77–88	Third floor of building in Paris area, winter and summer	Kirchner et al. (2002)
Houses, reduction compared with ratio with/without HEPA filtration unit	33	Australia, 15-week measurement, April (reduction is 40% for PM$_{10}$)	White et al. (2002)
Sulphur dioxide	Average ranges of I/O%: HVAC1= 1–10 and HVAC2/NV= 10–50		
Building, NV	40	Birmingham, UK, February 1996	Kukadia et al. (1996)
Building, HVAC system, particle filter	40	Birmingham, UK, February 1996	Kukadia et al. (1996)
Museums, well sealed, NV, N = 0.3/h	19	Bethnal Green Museum, UK	Cassar et al. (1999)
Museum with HVAC and gas filters, N = 1.3/h, 80% recirculation	14	Museum of London, UK	Cassar et al. (1999)
Office, NV, N = 1.6/h	40		Cassar et al. (1999)
Office, HVAC without recirculation or carbon filters, N = 1.2/h	40		Cassar et al. (1999)
Leaky room without portable carbon filter	45		Blades et al. (2000)
Leaky room with portable carbon filter	<LD		Blades et al. (2000)
Archive, storeroom, HVAC with fine dust filters	<20	1-week avg., The Hague, summer	Lanting (1990)
Archive, storeroom, HVAC with fine dust filters	8	1-week avg., The Hague, winter	Lanting (1990)
Archive, storeroom, HVAC with fine dust filters, 20% recirculation	22	The Hague	Lanting (1990)
Archive, storeroom, HVAC with fine dust filters, 35% recirculation	10	The Hague	Lanting (1990)
Archive, storeroom, HVAC with fine dust filters, 62% recirculation	5	The Hague	Lanting (1990)
Archive, storeroom, HVAC with fine dust filters, 93% recirculation	3	The Hague	Lanting (1990)
Museum, exhibit room with tapestry, NV	<20	1-week avg., Middelburg, The Netherlands, summer	Lanting (1990)
Museum, exhibit room with tapestry, NV	12	1-week avg., Middelburg, The Netherlands, winter	Lanting (1990)

Appendix 5. Permeability Coefficients of Various Products to Specific Gases

(unit: cm^3 (STP) cm cm^{-2} s^{-1} Pa^{-1} 10^{13})[a]

Products	Oxygen at 25°C	Water at 25°C	H_2S at 20°C	Ammonia at 20°C
Thin sheets (for bags, envelopes, encapsulation)				
Plastic laminated aluminum foil (such as Marvelseal)	0	0	0	0
Ethylene vinyl alcohol (EVOH)	0.0002	—	—	—
Poly(vinylidene chloride) (PVDC, Saran)	0.0038 [b]	0.027	7.0 [b]	—
Regenerated cellulose (Cellophane, cellulose hydrate)	0.0067	19 000	0.43	130 [c]
Polyamide 6, 6 (Dartek, nylon 6,6)	0.029 [b]	210 [b]	—	—
Poly(ethylene terephthalate) (PET, Mylar A, Hostaphan)	0.03	98	0.14	1.1
Polyethylene, high density (0.964 g cm^{-3}) (HDPE)	0.3	9	6.5	8
Cellulose acetate, plasticized	0.59 [b]	5500 [b]	4.6 [b]	—
Polypropylene, low density (0.907), crystallinity 50%	1.7 [b]	51 [b]	2.4	6.9
Polyethylene, low density (0.914) (LDPE)	2.2	68	27	21
Thick sheets (for boxes, windows, supports)				
Glass or metal	0	0	0	0
Polyamide 6, 6 (Dartek, nylon 6,6)	0.029 [b]	210 [b]	—	—
Poly(vinyl chloride), unplasticized (PVC)	0.034	210	0.14	3.7
Poly(methyl methacrylate)	0.12 [d]	480	—	—
Polyethylene, high density (0.964) (HDPE)	0.3	9	6.5	8
Polycarbonate (Lexan)	1.1 [b]	1100 [b]	—	—
Polypropylene, low density (0.907), crystallinity 50%	1.7 [b]	51 [b]	2.4	6.9
Polystyrene, biaxially oriented	1.9	840	—	—
Polyethylene, low density (0.914) (LDPE)	2.2	68	27	21
Poly(chloroprene) (Neoprene G)	3	840	—	—
Polytetrafluoethylene (PFTE, Teflon)	3.2	13	—	—
Acrylonitrile-butadiene-styrene (ABS)	4	—	—	—

a: Data mainly from Pauly (1989); STP = standard temperature (273.15K) and pressure (1.013 x 10^5 Pa).
The smaller the coefficient value, the better the impermeability; and, as indicated by the unit, the thicker the film or panel, the better the impermeability. These data do not include the possibility of a pin hole in the film and do not cover the permeability of foam products. Laminated or composite barrier films will have, at least, the permeability of the best single barrier film; see also permeability of plastic films to oxygen from Ankersmit et al. (2000).

b: Permeability determined at 30°C.

c: Permeability determined at 25°C.

d: Permeability determined at 34°C.

APPENDIX 6. DEPOSITION VELOCITY OF VARIOUS POLLUTANTS ON DIFFERENT SURFACES

Airborne pollutants	Products	Deposition velocity (m/h)	References
Nitrogen dioxide	Various products	0.01–4	Cass et al. (1989)
	Buffered paper (CaCO$_3$)	0.1	Lanting (1990)
	Cotton rag paper	0.1	Lanting (1990)
	Cotton fabric (bleached)	0.1	Lanting (1990)
Ozone	Plate glass	0.02	Sabersky et al. (1973)
	Various products	0.02–4	Cass et al. (1989); Kleno et al. (2001)
	Ash (untreated)	0.03	Kleno et al. (2001)
	Varnished ash	0.3	Kleno et al. (2001)
	Linoleum	0.1–0.3	Kleno et al. (2001)
	Stainless steel	0.3–0.5	Cass et al. (1989); Kleno et al. (2001)
	Polyethylene sheet	0.4	Sabersky et al. (1973)
	Carpets (nylon, olefin)	1–30	Kleno et al. (2001)
	Painted gypsum	1–20	Kleno et al. (2001)
	Various indoors surfaces	1.8 (average)	Lee et al. (1999)
	Cotton rag paper	2	Lanting (1990)
	Cotton fabric (bleached)	7	Lanting (1990)
	Unpainted gypsum	30	Kleno et al. (2001)
	White paper	>50	Kleno et al. (2001)
Particles, 0.1 μm[a]	Sea and ground surfaces	0.4	Slinn et al. (1978)
Particles, 1.0 μm	Sea and ground surfaces	0.3	Slinn et al. (1978)
Particles, 10 μm	Sea and ground surfaces	40	Slinn et al. (1978)
Sulphur dioxide	Cotton fabric (bleached)	0.3	Lanting (1990)
	Wool carpets	0.3–0.7	Crawshaw (1978)
	Buffered paper (CaCO$_3$)	0.7	Lanting (1990)
	Cotton rag paper	0.7	Lanting (1990)
	Wallpapers	6–20	Spedding and Rowlands (1970)
	Activated charcoal	30	Payrissat and Beilke (1975)
	Cements	60–90	Judeiki and Stewart (1976)

a: See Figure 7 for the deposition velocity as a function of particle diameter size.

Appendix 7. Methods for Measuring Air Leakage

Tracer gas decay test[a]	Tracer gas	Comments	References
	Carbon dioxide (CO_2)	Hole required.	Brimblecombe and Ramer (1983)
	Carbon dioxide (CO_2)	No hole needed, a commercial product.	Keepsafe (2002a)
	Dinitrogen oxide (N_2O)	Hole required.	Cassar and Martin (1994)
	Oxygen (O_2)	Hole required.	Brimblecombe and Ramer (1983); Daniel and Maekawa (1992)
	Sulphur hexafluoride (SF_6)	Commonly used for building, hole required.	ASHRAE (2001)
	Water (H_2O)	No hole required, problem of high moisture absorption capacity of enclosure products.	Daniel and Maekawa (1992); Padfield (1966); Thomson (1977)

Air pressurization test	Comments	References
	Values based on average of positive and negative pressure tests.	Cassar and Martin (1994); Padfield (1966); Toishi and Koyano (1988); Michalski (1994b)

Theoretical prediction	Comments	References
	Simplified tables and charts of leakage are given for 0.1-, 1-, and 10-m^3 enclosure, as a function of crack width, hole width, and wall permeability.	Michalski (1994b)

a: Results should be normalized to a reference gas or gas mixture such as air (78% nitrogen, 21% oxygen, and 1% argon); otherwise the tracer gas should be specified. Values should be expressed in units of air exchange per day (1/day or day^{-1}).

Appendix 8. Guidelines for Light Intensities

Appendix 8A. Guidelines for light intensities for museum, gallery, library, and archive collections

Category	LOAED [a]	Preservation targets [b]		
		1000 yrs	**100 yrs**	**10 yrs**
High sensitivity ISO 1, 2, 3	ISO 2: 1.0 Mlx h (million lux hour)	50 lx for 20 h/yr	50 lx for 25 days/yr 500 lx for 25 h/yr [c]	50 lx for 250 days/yr 500 lx for 25 days/yr
Medium sensitivity ISO 4, 5, 6	ISO 4: 10 Mlx h	50 lx for 25 days/yr 500 lx for 20 h/yr	50 lx for 250 days/yr 500 lx for 25 days/yr	340 lx for 365 days/yr 500 lx for 250 days/yr
Low sensitivity ISO 7, 8, above	ISO 7: 300 Mlx h	100 lx for 365 days/yr 500 lx for 75 days/yr	1000 lx for 365 days/yr [d] (500 lx/yr for target 200 yrs)	

a: LOAED based on grey scale 4 (British Standard BS1006) (Michalski 1987). Some ISO blue wool standard equivalencies for colorants and materials are available in Appendix 8B.

b: Number of years before observing a low adverse effect. The illuminations allowed are based on 8 h of light exposure per day, with UV radiation filtered. Intermediate preservation targets can be used as well. Variations in light intensity versus exposure period are also possible (50 lx for 300 days/yr = 100 lx for 150 days/yr).

c: If the object has dark surfaces or low contrast detail, the observers are old, or a difficult task must be performed, light levels may be increased up to 10 times: about 500 lx (2 of 4 visibility factors) (Michalski 1997).

d: The practice of having a light intensity higher than 500 lx may encourage other uninformed museums to use unnecessarily high light levels without doing a proper light fading risk assessment of the collection.

Half-faded colorants will have their LOAED shift up by an order of magnitude (e.g. a half-faded colorant ISO 2 can be shifted up to ISO 4). It takes about 10 cumulative adverse effects to cause 50% fading (it will take 100 yrs for a colorant with a preservation target of 10 yrs to become half faded) and a 90% fade should happen after 18 or 20 adverse effects (a colorant with a target of 10 yrs will be almost completely faded in about 500 yrs). If an illumination of 50 lx is applied to all objects, 8 h per day, every day, high sensitivity colorants will be 50% faded after 20 yrs, medium sensitivity colorants after 200 yrs, and low sensitivity colorants after 6000 yrs.

Appendix 8B. ISO blue wool standard equivalencies for colorants and materials

Category	LOAED	Materials (Michalski 1997, 1999)[a]
High sensitivity ISO 1, 2, 3	ISO 2: 1.0 Mlx h (million lux hour)	- Most plant extracts, hence most historic bright dyes and lake pigments in all media: yellows, oranges, greens, purples, many reds, blues. - Insect extracts, such as lac (yellow), cochineal (carmine) in all media. - Most early synthetic colours such as the anilines, all media. - Many cheap synthetic colorants in all media. - Most felt tip pens including blacks. - Most dyes used for tinting paper in the 20th century. - Most colour prints with "colour" in the name.
Medium sensitivity ISO 4, 5, 6	ISO 4: 10 Mlx h	- A few historic plant extracts, particularly alizarin (madder red) as a dye on wool or as a lake pigment in all media. It varies throughout the range of media and can reach into the low category, depending on concentration, substrate, and mordant. - The colour of most furs and feathers. - Most colour prints with "chrome" in the name (silver dye bleach prints).
Low sensitivity ISO 7, 8, above	ISO 7: 300 Mlx h	- Artists' palettes classified as "permanent" (a mix of truly permanent and low light sensitivity paints, e.g. ASTM D4303 Category I; Winsor and Newton AA). - Structural colours in insects (if UV blocked). - A few historic plant extracts, especially indigo on wool. - Silver/gelatine black-and-white prints, not RC paper, and only if all UV blocked. - Many high quality modern pigments developed for exterior use, automobiles.

a: See also Ashley-Smith et al. (2002) for other categorizations of ISO blue wool standard equivalencies.

GLOSSARY

The following terms are defined on the basis of the way in which they are used in this book and/or the field of preservation. Some terms may have wider or alternate meanings in other fields.

Accelerated ageing (accelerated testing)
A thermal deterioration experiment on materials that provides an estimation of their loss of physical or chemical properties over an extended period of time.

Adverse effect
The first visually perceptible change; a specific chemical or physical characteristic of the material/ object usually considered abnormal or undesirable. This term is commonly used for risk analysis in the fields of health, safety, and environmental policy.

Aerodynamic diameter
The diameter of a sphere with a unit density that has an equivalent aerodynamic behaviour to that of the particle in question; an expression of aerodynamic behaviour of an irregularly shaped particle in terms of the diameter of an idealized particle. Particles with different dimensions and shapes may have the same aerodynamic diameter.

Aerosol
A suspension of fine solid or liquid particles in air or gas that forms a mist, fog, or smoke; a colloidal system consisting of a gas (usually air) continuous medium in which particles of solid or liquid are dispersed.

Agent of deterioration
The active threats to museum collections can be grouped into nine different agents of deterioration: direct physical forces; thieves (including vandals and displacers); fire; water; pests (insects, vermin, mould, etc.); pollutants; radiation; incorrect temperature; and incorrect RH.

Air exchange rate
The rate at which outside air replaces inside air in a given space (a room or an enclosure).

Airborne pollutant
A gas, vapour, liquid, or solid particle (of either anthropogenic or natural origin) carried by the air that is known to cause adverse effects to cultural property. Implicitly, "airborne" refers to the mechanism for reaching the object. See also *Adverse effect* and *Pollutant*.

Ambient air
The air surrounding an object. In other fields this term is often used to refer to the outdoor air or, more specifically, the air subjected to meteorological and climatic changes and pollution from combustion due to automobiles, trucks, power plants, and incinerators.

Anoxic environment
An environment that has low levels of oxygen; anaerobic, oxygen-reduced, and oxygen-free environments are synonyms.

Anthropogenic
The impact of humans on the environment; induced or altered by the presence or activities of humans.

Carbonyl
A compound made up of one atom of carbon and one atom of oxygen connected by a double bound ($C=O$); found in aldehydes, ketones, carboxylic acids, and derivatives. The term includes some chemical groups (e.g. ketones and polypeptides) that are not known to cause adverse effects on objects, so its use should be avoided when discussing harmful pollutants.

Catalyst
A substance used in small proportion that augments the rate of a chemical reaction without (in theory) being chemically changed at the end of the reaction.

Cellulose-based material
A material (e.g. paper, rayon, and cellulose lacquers and films) that is manufactured from cellulose. Cellulose is a complex carbohydrate that is the chief constituent of the walls of plant cells. The main sources of cellulose are wood, cotton, and other fibrous materials.

Collection
A set of objects that have been brought together on the basis of common characteristics or are to be treated as a group. Any set of objects in the same location can be called, to some extent, a collection.

Control strategy
A co-ordinated combination of measures (e.g. avoid, block, dilute, filter/sorb, reduce reactions, and reduce exposure time) to reduce one specific or many agents of deterioration present at a certain level or to minimize their adverse effects on objects. These measures can be based on predefined specifications.

Cost–benefit analysis

An analytical method designed to evaluate different scenarios in terms of their likely cost and potential benefit. It contains key steps such as the development of weighted criteria, the evaluation of different options (control strategies) against the criteria, and the assessment of possible adverse consequence of the tentative options. This tool facilitates the decision-making process.

Crustal particles

Particles emitted directly from a non-industrial surface (e.g. paved roads and traffic, construction, agricultural operations, high wind events) and some industrial processes.

Damage

See *Loss of value*.

Deposition velocity (or mass transfer coefficient)

A measure of the rate at which a pollutant reaches the surface of a material. A common unit is metres per second (m/s) or per hour (m/h). Visualize the time it takes a particle to fall a distance of 1 m. If it takes 2 h, the particle has a deposition velocity of 0.5 m/h (1 m/2 h). This number is determined experimentally and is used as a constant to simplify modelling. The deposition velocity refers only to mass transfer from the ambient air to a surface. The mass transfer coefficient refers to mass transfer from the ambient air to a surface, or from a surface to the ambient air. Its value will be positive or negative depending on the direction of the transfer.

Deterioration or degradation

A change in the material state of an object.

Dose

The concentration of a pollutant multiplied by the exposure time. In toxicology, dose refers to the amount (weight or volume) of a chemical or biological agent absorbed or inhaled by an individual. See also *LOAED*.

Electrostatic precipitator

A type of air pollution control system that uses high voltage fields to give particulate matter an electrical charge. When the charged particles approach an electrically grounded collection plate they accumulate as a dust layer, which is mechanically removed (at least partially) on a routine basis.

Enclosure

A collection of products that surrounds a limited space (e.g. a plastic bag, display case, storage cabinet, or transportation box).

Environmental performance

See *Performance target*.

Equilibrium concentration

The maximum concentration a vapour emitted from a material can reach in an absolutely airtight enclosure; the emission and sorption rate of the vapour are equal. Also referred to as saturation concentration.

Fatty acid

A carboxylic acid containing a long hydrocarbon chain with no double bonds (saturated fatty acid), one double bond (monounsaturated fatty acid), or two or more double bonds (polyunsaturated fatty acid) derived from animal or vegetable fats and oils.

Fixed-bed filter

A filter containing multi-layered fine size sorbents that are tightly bonded to a three-dimensional network of adhering fibres on non-woven carriers or to open cell polyurethane foam structures (see Figure 32).

HVAC system

A heating, ventilating, and air-conditioning system that includes any interior surface of the facility's air distribution system for conditioned spaces and/or occupied zones.

Hydrolysis

From Greek, hydro = water and lysis = to break or separate. A chemical process in which scission of a chemical bond occurs via a reaction with water. The degradation weakens or breaks molecular bonds, thereby leading to embrittlement and discoloration. Specific hydrolysis processes may be catalysed by acids, alkalis, or enzymes according to the type of reaction. The polyester chemical bonds in tape binder polymers are subject to hydrolysis, producing alcohol and acid end groups. For other organic materials highly susceptible to hydrolysis, see "water vapour" in Table 2.

Institutional preservation policy

See *Preservation policy*.

Investigation

A process conducted to prevent damage or determine cause(s) of damage on objects. It includes gathering and analysing information about the physical and chemical environments, the human activities, and the nature of the objects. Adequate recommendations should be made when appropriate.

Key airborne pollutants

Pollutants whose sources, reactivities, and permissible concentrations are equal to or less than the other pollutants of the same chemical group. In preservation, the key pollutants are acetic acid, hydrogen sulphide, nitrogen dioxide, ozone, fine particles, sulphur dioxide, and water vapour. See also *Airborne pollutant*.

Linear reciprocity principle

See *Reciprocity principle*.

Liquid product

A building or enclosure product with high liquid content characterized by relatively high water and/or VOC emissions during and immediately after installation followed by much lower emissions after drying is complete (e.g. adhesives, cleaning products, coatings, floor waxes, spray polishes, and silicone caulks).

LOAED (lowest observed adverse effect dose)

The cumulative dose of a pollutant (concentration x time) at which the first signs of adverse effects are observed (measured) on a material. When a NOAEL (no observed adverse effect level) cannot be determined with confidence or when it is not feasible, a dose can be determined as the product of concentration of the pollutant and the time required to observe the first signs of an adverse effect. See *Adverse effect*.

Loss of value

The decrease in attribution or value (aesthetic, scientific, historic, symbolic, monetary, spiritual, etc.) of the object. The notion of loss of value is subjective, i.e. it can vary depending on the person and the time.

Mass transfer coefficient

See *Deposition velocity*.

Material

A substance that composes an object or a product, e.g. copper, oak, cotton are materials. See also *Object* and *Product*.

Museum

A building structure housing objects or collections, e.g. galleries, archives, libraries, churches, historical houses, and private houses.

NOAEL (no observed adverse effect level)

The highest level of a pollutant that does not produce an adverse effect on a specific chemical or physical characteristic of a material in a specific experimental set-up (analytical method, exposure time, temperature, RH, etc.). Some effects may be produced at this level, but they are not considered adverse, nor precursors to adverse effects. See *Adverse effect*.

Nucleation

A process by which gases or fine particle compounds interact and form aggregates with the same molecules (homogeneous) or with different molecules (heterogeneous).

Object (or cultural property)

An item judged by society, or by some of its members, to be of historical, artistic, social, or scientific importance. Objects can be composed of one or more materials, and can be movable (such as works of art, artifacts, books, archival-related items, and items of natural, historical, or archaeological origin) or immovable (such as architectural interiors or structures of historical or artistic interest).

Order of magnitude

A 10-fold difference between two quantities.

Organic object or material

Object or material made of compounds containing carbon atoms, with the exception of carbon dioxide and carbon monoxide. Organic materials can be natural or man-made. This term originally referred to materials derived from living things (i.e. from plants and animals).

Oxidant

A qualitative term that includes any and all trace gases that have a greater oxidation potential than oxygen, e.g. ozone, chlorine, peroxyacetyl nitrate, peroxides, sulphuric acid, and nitric acid.

Oxidation

A reaction in which oxygen combines chemically with another substance. Usage of this term has been broadened to include any reaction in which electrons are transferred.

Oxidized sulphur gases (S^+)

Sulphur compounds that have the sulphur atom in the oxidation state of +4 or +6, e.g. SO_2 and H_2SO_4, respectively.

Pareto's principle
The 80-20 rule. Vilfredo Pareto (1848–1923) was an Italian economist who, in 1906, observed that 20% of the Italian people owned 80% of their country's accumulated wealth. Over time and through application in a variety of environments, this has come to be called Pareto's principle or the 80-20 rule. This 80-20 mix reminds us that the relationship between input and output is not balanced. In the context of management, this rule of thumb is a useful heuristic that can be applied when there is a question of the effectiveness of increasing effort, expense, or time versus diminishing returns.

Performance target
The goal or focus of performance based on maximum average concentration of one or a set of pollutants surrounding collections in the room or enclosure. It can be obtained through quantitative or semi-quantitative monitoring methods, or from a set of specifications to be fulfilled.

Photo-oxidation
Oxidation reactions induced by radiant energy (UV or visible), e.g. the loss of one or more electrons from a chemical species as a result of photo-excitation of that species, and the reaction of a substance with oxygen under the influence of radiation. When oxygen is involved, the process is also called photo-oxygenation. Reactions in which neither the substrate nor the oxygen are electronically excited are sometimes called photo-initiated oxidations.

Photo-reduction
Reduction reactions induced by radiant energy, e.g. the addition of one or more electrons to a photo-excited species, and the photochemical hydrogenation of a substance.

Physical attrition
The act or process of rubbing together or wearing down by friction; specifically, the mutual wear and tear (due to rubbing, grinding, knocking, scraping, and bumping against one another) that loose rock fragments or particles undergo while being moved about by wind, with a resulting reduction in size and increase in roundness.

Pollutant
A gas, vapour, liquid, or solid particle of either anthropogenic or natural origin that is known to cause adverse effects to cultural property. The adverse effects are particularly associated with

chemical reactions between the pollutant and one or more components of the object. They also include the deposition or migration of compounds that can cause adverse effects on the object (see *Adverse effect*). There are three mechanisms for a pollutant to reach an object: it can be carried by the air (airborne pollutant); it can be transferred from one material to another at a point of contact (the material from which the pollutant originates can be either an object or a product); or it can be present as part of the object. In this last mechanism, no mass transfer occurs because the pollutant is already present as an intrinsic part of the original object or it is generated by an internal degradation process.

Pollutant–material system
The interrelationship or interaction between a specific pollutant and a specific material (which can be either the whole object or a part of it).

Pollutant–object system
Same as pollutant–material system, except that the emphasis is on the museologic character of the object.

Preservation assessment
An evaluation of the degree of protection of an object or a collection. The preservation assessment relies on risk assessment tools (NOAEL and LOAED) and is usually expressed in terms of the risk or rate of deterioration caused by agents of deterioration such as pollutants. See *Risk assessment*.

Preservation policy
The institutional goals and principles in relation to overall performance in collection preservation. The policy provides a framework for actions.

Preservation target
A goal or focus on the degree of preservation of objects; the period of time over which the object or collection should have no or minimum risk of deterioration. Minimum risk of deterioration is determined by using the concept of LOAED. This approach allows for the extrapolation of the maximum concentration of pollutants over the time period to ensure the minimum risk of deterioration. In a few cases, NOAEL can be used if the data are available and this approach is feasible. NOAEL represents the maximum concentration of a pollutant where no adverse effect is observable on an object. In this case, the preservation target becomes zero damage over an undetermined exposure time. See also *LOAED* and *NOAEL*.

Primary pollutants
Pollutants emitted directly from a source, e.g. SO_2 and NO from combustion, ash from volcanic eruption, and acetic acid from wood. See also *Secondary pollutants*.

Product
A manufactured or processed substance made up of one or more materials that is used in the care and housing of collections or objects. For example, plywood is a product made of the materials wood and adhesive. Specifications for selecting products use a variety of terms such as inert, stable, safe (no or low risk), or suitable. These terms are not standardized, but "inert" almost always refers to a very high level of stability that is not usually mandatory in conservation.

Reciprocity principle
The concept that as long as the dose (concentration of a pollutant x exposure time) is constant, the adverse effect will be the same, i.e. doses of 450 μg m^{-3} for 1 yr, 45.0 μg m^{-3} for 10 yrs, and 4.50 μg m^{-3} for 100 yrs should all produce the same adverse effect. This is analogous to the relationship between light intensity and shutter speed in photography, where 1/25 s at f11 produces the same light density as 1/100 s at f5.6. In extreme conditions, linearity is not always applicable.

Reduced sulphur gases (S⁻)
Sulphur compounds that have the sulphur atom in the oxidation state of -2 [e.g. hydrogen sulphide (H_2S), carbon disulphide (CS_2), carbonyl sulphide (OCS)] or -1 [e.g. dimethyl disulphide (CH_3SSCH_3)].

Relative humidity (RH)
The ratio of the amount of water vapour actually present in a given volume of air and the amount that would be required to saturate that air at a given temperature, expressed as a percentage. For example, the saturation value of air at 20°C is 17.30 g m^{-3}; if the actual amount of water vapour present is 8.65 g m^{-3}, the RH is 50%.

Risk assessment
An evaluation of agents of deterioration in the environment and their potential adverse effects on the collection. Micro-scale risk assessment looks at the adverse effects of a particular pollutant on a particular material, and can be done by comparing the LOAED or NOAEL value of the specific pollutant–material system with the concentration of the pollutant in the surrounding environment of the object.

Macro-scale risk assessment can be done for a mixed collection by looking at the adverse effects of the key airborne pollutants on the most sensitive or most representative materials. See also *Preservation assessment*.

Secondary pollutants
Pollutants not emitted directly from a source but rather formed in the air by the reactions or interactions of primary pollutants. One example is ozone, which is one of many secondary pollutants that make up photochemical smog. Note that some pollutants may be both primary and secondary (i.e. they can be emitted directly or formed from other primary pollutants).

Smog
A compression of the words "smoke" and "fog." The term is used to designate pollution problems involving reduced visibility. Smog can be subdivided into acid smog which is associated with SO_2 and particulate matter, and photochemical smog which results from increased concentrations of ozone (its principal component) when precursor pollutants (VOCs and NO_X) react in the presence of ultraviolet radiation. Acid smog usually happens in cold weather due to meteorological conditions and the high use of heating systems based on coal combustion. With the reduction in SO_2 during the last decades, the major pollutant of winter smog has become particulate matter.

Sorb
To take up and hold either by adsorption or absorption. See *Sorbent*.

Sorbent
A material that provides a sorption function. Sorbents can extract some compounds present in the ambient air and retain them by an affinity or reaction process. They can also be used to collect gases and vapours during air sampling. A sorbent may work through absorption (interactions taking place largely within the pores of solids) or adsorption (interactions taking place on solid surfaces). The processes involved can also be divided into chemisorption (chemical bonding with the substrate) and physisorption (physical attraction, such as weak electrostatic forces). Desorption processes can also happen if the sorbed gases have not already reacted and/or the level of pollutants decreases in the ambient air. In conservation, the term "scavenger" is often used as a synonym for sorbent.

Sorptive material
See *Sorbent*.

Spatial gradient
The rate of increase or decrease of a physical or chemical value through space.

Specification
An accurate description of the technical requirements for the performance of building features, portable fittings, or procedures.

SPME (solid-phase microextraction)
A process developed to facilitate rapid sample preparation both in the laboratory and in situ. A small amount of the extracting phase associated with a solid support is placed in the environment of interest for a predetermined time. SPME is compatible with analytic separation/detection by gas chromatography or high-performance liquid chromatography.

Stack pressure
The pressure difference that results when a temperature difference creates air movement within an enclosure or duct.

Steady-state concentration
The concentration of airborne compounds in an enclosure that results from the interaction of all different parameters involved in the transfer of compounds (e.g. emission rate, sorption rate, and air exchange rate).

Suspension time
The average period for which an emitted or resuspended pollutant stays in the ambient air or is carried in the atmosphere without chemical conversion. Also referred to as atmospheric residence time.

Synergistic effect
A phenomenon in which the effect obtained from the combined action of two distinct compounds is greater than that obtained when their independent actions are added together. The opposite phenomenon is the antagonist effect in which the effect obtained is less than predicted by adding their independent actions.

Volatile organic compounds (VOCs)
A class of chemical mixtures that contain one or more carbon atoms, and exist in vapour form at room temperature (i.e. with lower boiling point limits between 50 and 100°C, and upper limits between 240 and 260°C). This definition is based on the methods used to sample VOCs. The term is generally applied to organic solvents, certain paint additives, aerosol spray can propellants, fuels (such as gasoline and kerosene), petroleum distillates, dry cleaning products, and many other industrial and consumer products ranging from office supplies to building materials. VOCs are also naturally emitted by a number of plants and trees. This term should be avoided when discussing preservation of cultural property as many VOCs are not known to cause adverse effects on objects. In fact, most VOCs (except amines, aldehydes, and carboxylic acids) are considered harmless in a normal museum environment. TVOC refers to total VOCs.

REFERENCES

Note: All Web pages were accessed in December 2003.

ACS (American Chemical Society) and Resource for the Future. *Understanding Risk Analysis: A Short Guide for Health, Safety, and Environmental Policy Making.* Internet edition (1998), pp. 6–27. <www.rff.org/misc_docs/risk_book.pdf>

Adams, S., P. Brimblecombe, and Y.H. Yoon. "Comparison of Two Methods for Measuring the Deposition of Indoor Dust." *The Conservator* 25 (2001), pp. 90–94.

Adelstein, P.Z., J.M. Reilly, D.W. Nishimura, and K.M. Cupriks. "Hydrogen Peroxide Test to Evaluate Redox Blemish Formation on Processed Microfilm." *Journal of Imaging Technology* 17 (1991), pp. 91–98.

Adelstein, P.Z., J.M. Reilly, D.W. Nishimura, C.J. Erbland, and J.L. Bigourdan. "Stability of Cellulose Ester Base Photographic Film: Part V - Recent Findings." *SMPTE Journal* (1995), pp. 439–447.

Aero-Laser. *Aero-Laser GMBH.* nd. <www.aero-laser.de/> (Note: Look under Unisearch products for LOZ-3 and LMA-3D detectors, look under Aero-Laser GMBH products for AL-4021.)

Air Pollution Training Institute. "Control Techniques." *Module 6: Air Pollutants and Control Techniques, Basic Concepts in Environmental Sciences.* 2000. <www.epin.ncsu.edu/apti/ol_2000/module6/mod6fram.htm>

Alberta Environmental Protection. *Air Quality in Alberta; April to June 1998.* 1998. <www3.gov.ab.ca/env/air/airqual/98AprJun.rep.pdf>

Allegrini, I., and F. De Santis. "Measurement of Atmospheric Pollutants Relevant to Dry Acid Deposition." *Critical Reviews in Analytical Chemistry* 21 (1989), pp. 237–255.

Ankersmit, H.A. Netherlands Institute for Cultural Heritage. Personal communication, 2001.

Ankersmit, H.A., G. Noble, L. Rodge, D. Stirling, N.H. Tennent, and S. Watts. "The Protection of Silver Collections from Tarnishing." pp. 7–13 in *Tradition and Innovation: Advances in Conservation* (edited by A. Roy and P. Smith). London: International Institute for Conservation, 2000.

APT/AIC (Association for Preservation Technology International/American Institute for Conservation of Historic and Artistic Works). "New Orleans Charter for Joint Preservation of Historic Structures and Artifacts." *APT Communique* 21, 2 (1992). <palimpsest.stanford.edu/bytopic/ethics/neworlea.html>

Arizona Instrument. *Jerome 632-X Hydrogen Sulfide Analyzer.* nd. <www.azic.com/products_631.aspx>

Arney, J.S., A.J. Jacobs, and R. Newman. "The Influence of Oxygen on the Fading of Organic Colorants." *Journal of AIC* 18 (1979), pp. 108–117.

Arrhenius, S. "Uber die Reaktionsgeschwindigkeit' bei der Inversion von Rohzucker durch Sauren." *Zeitschrift für Physikalische Chemie* 4 (1889), pp. 226–248.

Ashley-Smith, J. *Risk Assessment for Object Conservation.* Oxford: Butterworth Heinemann, 1999, pp. 72–119.

Ashley-Smith, J., A. Derbyshire, and B. Pretzel. "The Continuing Development of a Practical Lighting Policy for Works of Art on Paper and Other Object Types at the Victoria and Albert Museum." pp. 3–8 in *Preprints of ICOM-CC Conference, Volume 1* (edited by R. Vontobel). London: James & James Science Publishers Ltd., 2002.

ASHRAE. "Museums, Libraries, and Archives." Chapter 21 in *Heating, Ventilating, and Air-Conditioning: Applications.* ASHRAE Handbook. Atlanta, 2003.

ASHRAE (American Society of Heating, Refrigerating and Air-Conditioning Engineers). *Ventilation for Acceptable Indoor Air Quality.* ASHRAE Standard 62-2001. Atlanta, 2001a.

ASHRAE. "Ventilation and Infiltration." Chapter 26 (pp. 26.26–26.27) in *Heating, Ventilating, and Air-Conditioning: Fundamentals*. ASHRAE Handbook. Atlanta, 2001b.

ASHRAE. "Air Contaminants." Chapter 12 in *Heating, Ventilating, and Air-Conditioning: Fundamentals*. ASHRAE Handbook. Atlanta, 2001c.

ASHRAE. "Air Cleaners for Particulate Contaminants." Chapter 25 (pp. 25.3–25.5) in *Heating, Ventilating, and Air-Conditioning: Systems and Equipment*. ASHRAE Handbook. Atlanta, 2001d.

ASHRAE. *Methods of Testing General Ventilation Air-Cleaning Devices for Removal Efficiency by Particle Size*. ANSI/ASHRAE Standard 52.2-1999, 1999a, pp. 35–39.

ASHRAE. "Control of Gaseous Indoor Air Contaminants." Chapter 44 (pp. 44.1–44.21) in *Heating, Ventilating, and Air-Conditioning: Applications*. ASHRAE Handbook. Atlanta, 1999b.

ASTM (American Society for Testing and Materials). "Standard Practice for Planning the Sampling of the Ambient Atmosphere." ASTM D1357-95. pp. 15–18 in *Water and Environmental Technology*. Annual Book of ASTM Standards. Volume 11.03. West Conshohocken, PA, 2001a.

ASTM. "Standard Practice for Sampling Workplace Atmospheres to Collect Gases or Vapors with Solid Sorbent Diffusive Samplers." ASTM D4597-95. pp. 289–292 in *Water and Environmental Technology*. Annual Book of ASTM Standards. Volume 11.03. West Conshohocken, PA, 2001b.

ASTM. "Standard Test Method for Measurement of Formaldehyde in Indoor Air (Passive Sampler Methodology)." ASTM D5014-94. pp. 373–378 in *Water and Environmental Technology*. Annual Book of ASTM Standards. Volume 11.03. West Conshohocken, 2001c.

ASTM. "Standard Test Method for Measurement of Nitrogen Dioxide Content of the Atmosphere (Griess-Saltzman Reaction)." ASTM D1607-91. pp. 19–23 in *Water and Environmental Technology*. Annual Book of ASTM Standards. Volume 11.03. West Conshohocken, 2001d.

Bacci, M., M. Picollo, S. Porcinai, and B. Radicati. "Tempera-painted Dosimeters for Environmental Indoor Monitoring: A Spectroscopic and Chemometric Approach." *Environmental Science & Technology* 34 (2000), pp. 2859–2865.

Backlund, P., B. Fjellstrom, S. Hammarback, and B. Maijgren. "The Influence of Humidity on the Reaction of Hydrogen Sulphide with Copper and Silver." *Arkiv for Kemi* 26 (1966), pp. 267–277.

Bastidas, J.M., A. López-Delgado, E. Cano, J.L. Polo, and F.A. López. "Copper Corrosion Mechanism in the Presence of Formic Acid Vapour for Short Exposure Times." *Journal of the Electrochemical Society* 147 (2000), pp. 999–1005.

Bayer, C. "Fungal Growth May Contribute to VOC Emissions in Indoor Air." *Indoor Air Quality* (December 1993), pp. 6–7.

Bégin, P., S. Deschatelets, D. Grattan, N. Gurnagul, J. Iraci, E. Kaminska, D. Wood, and X. Zou. "The Effect of Air Pollutants on Paper Stability." *Restaurator* 20 (1999), pp. 1–21.

Bellan, L.M., L.G. Salmon, and G.R. Cass. "A Study on the Human Ability to Detect Soot Deposition onto Works of Art." *Environmental Science & Technology* 34 (2000), pp. 1946–1952.

Bennett, H.E., R.L. Peck, D.K. Burge, and J.M. Bennett. "Formation and Growth of Tarnish on Evaporated Silver Films." *Journal of Applied Physics* 40 (1969), pp. 3351–3360.

Bernard, N.L., C.M. Astre, B. Vuillot, M.J. Saintot, and M. Gerber. "Measurement of Background Urban Nitrogen Dioxide Pollution Levels with Passive Samplers in Montpellier, France." *Journal of Exposure Analysis and Environmental Epidemiology* 7 (1997), pp. 165–178.

Bernard, N.L., M.J. Gerber, C.M. Astre, and M.J. Saintot. "Ozone Measurement with Passive Samplers: Validation and Use for Ozone Pollution Assessment in Montpellier, France." *Environmental Science & Technology* 33 (1999), pp. 1217–1222.

Biddle, W. "Art in Cornell Museum Coated by Chemical Used in Steam Lines." *The New York Times* (July 29, 1983), pp. B-1 and B-2.

Blades, N. "Nitrogen Dioxide and Sulfur Dioxide Measurement at the British Library and National Library of Wales, 1995-1996." pp. 23 and 31 in *Deterioration and Conservation of Vegetable Tanned Leather; Environment Leather*. Project EV5V-CT94-0514, Protection and Conservation of European Cultural Heritage Research Report Number 6. Copenhagen: L.P. Nielsen Offset Desktop Bogtryk, 1997.

Blades, N., T. Oreszczyn, B. Bordass, and M. Cassar. *Guidelines on Pollution Control in Museum Buildings.* London: Museums Association, 2000, pp. 20–21.

Bosworth, J. "Retrofitting Old Exhibit Cases: A Search for Economic and Safe Cabinetry." *Exhibitionist* 20 (2001), pp. 20–24.

Bower, J. "Air Filters." *East West* (September/October, 1991), p. 28. <www.hhinst.com/Artairfilters.html>

Bowser, D., and D. Fugler. "Indoor Ozone and Electronic Air Cleaners." pp. 670–675 in *Proceedings of the 9th International Conference on Indoor Air Quality and Climate, Volume 2* (edited by H. Levin). Santa Cruz: International Conference on Indoor Air Quality and Climate, 2002.

Brimblecombe, P. "Pollution Studies." pp. 25–30 in *Deterioration and Conservation of Vegetable Tanned Leather; Environment Leather.* Project EV5V-CT94-0514, Protection and Conservation of European Cultural Heritage Research Report Number 6. Copenhagen: L.P. Nielsen Offset Desktop Bogtryk, 1997.

Brimblecombe, P. "The Balance of Environmental Factors Attacking Artifacts." pp. 67–79 in *Durability and Change: The Science, Responsibility, and Cost of Sustaining Cultural Heritage* (edited by W.E. Krumbein, P. Brimblecombe, D.E. Cosgrove, and S. Staniforth). Toronto: John Wiley & Sons, 1994.

Brimblecombe, P., and B. Ramer. "Museum Display Cases and the Exchange of Water Vapour." *Studies in Conservation* 28 (1983), pp. 179–188.

Brimblecombe, P., D. Shooter, and A. Kaur. "Wool and Reduced Sulphur Gases in Museum Air." *Studies in Conservation* 37 (1992), pp. 53–60.

Brokerhof, A.W., and M. Van Bommel. "Deterioration of Calcareous Materials by Acetic Acid Vapour: A Model Study." pp. 769–775 in *Preprints of ICOM-CC Conference, Volume 2* (edited by J. Bridgland). London: James & James Science Publishers Ltd., 1996.

Brown, T. "Show-case Induced Cadmium Corrosion." Abstracts from the UK Conservation Science Group Meeting, 30 September 1998. <www.chemsoc.org/networks/csn/abstracts.htm>

Byne, L. StG. "The Corrosion of Shells in Cabinets." *Journal of Conchology* 9 (1899), pp. 172–178, 253–254.

Calmes, A. "Charters of Freedom of the United States." *Museum* 146 (1985), pp. 99–101.

Camuffo, D., R. Van Grieken, H.-J. Busse, G. Sturaro, A. Valentino, A. Bernardi, N. Blades, D. Shooter, K. Gysels, F. Deutsch, M. Wieser, O. Kim, and O.U. Ulrych. "Environmental Monitoring in Four European Museums." *Atmospheric Environment* 35 (Supplement 1) (2001), pp. S127–S140.

Caple, C. *Conservation Skills: Judgement, Methods and Decision Making.* New York: Routeledge, 2000, pp. 170–181.

Carroll, J.F., and J.M. Calhoun. "Effect of Nitrogen Oxide Gases on Processed Acetate Film." *Journal of the SMPTE* 64 (1955), pp. 501–507.

Cass, G.R., J.R. Druzik, D. Grosjean, W.W. Nazaroff, P.M. Whitmore, and C.L. Whittman. *Protection of Works of Art from Atmospheric Ozone.* Research in Conservation No. 5, pp. 34 and 70. Marina del Rey: The Getty Conservation Institute, 1989. <www.getty.edu/conservation/resources/ozone.pdf>

Cass, G.R., J.R. Druzik, D. Grosjean, W.W. Nazaroff, P.M. Whitmore, and C.L. Whittman. *Protection of Works of Art from Photochemical Smog.* Final report. Pasadena, CA: The Getty Conservation Institute, 1988, pp. 228–255.

Cassar, M., N. Blades, and T. Oreszczyn. "Air Pollution Levels in Air-conditioned and Naturally Ventilated Museums: A Pilot Study." pp. 31–37 in *Preprints of ICOM-CC Conference, Volume 1* (edited by J. Bridgland). London: James & James Science Publishers Ltd., 1999.

Cassar, M., and G. Martin. "The Environmental Performance of Museum Display Cases." pp. 171–173 in *Preventive Conservation: Practice, Theory and Research* (edited by A. Roy and P. Smith). London: International Institute for Conservation of Historic and Artistic Works, 1994.

Cavallo, D., D. Alcini, M. De Bortoli, D. Carrettoni, P. Carrer, M. Bersani, and M. Maroni. "Chemical Contamination of Indoor Air in Schools and Office Buildings in Milan, Italy." pp. 45–50 in *Proceedings of the 6th International Conference on Indoor Air Quality and Climate, Indoor Air '93, Volume 2* (edited by P. Kalliokoski, J. Jantunen, and O. Seppanen). Helsinki: International Conference on Indoor Air Quality and Climate, 1993.

CCI (Canadian Conservation Institute). *The Analytic Hierarchy Process Program*. 2003. <www.cci-icc.gc.ca/links/pollutants/index_e.shtml>

CEN (European Committee for Standardisation). *Particulate Air Filters for General Ventilation: Determination of the Filtration Performance*. EN 779. Brussels, 2002.

Cermakova, D., and Y. Vlchkova. "Metallic Corrosion in an Atmosphere Polluted with Formaldehyde." pp. 497–508 in *Proceedings of the Third International Congress on Metallic Corrosion*. Moscow, 1966.

Chahine, C. "Acidic Deterioration of Vegetable Tanned Leather." pp. 75–79 in *Leather: Its Composition and Changes with Time* (edited by C. Calnan and B. Haines). Northampton: The Leather Conservation Center, 1991.

Charcoal Cloth Limited. "Adsorptive Performance of ACC on Some Specific Vapours." Information from the company, 1990.

Chaumier, S. "Un traitement curatif de désinsectisation par anoxie sous atmosphere inerte : au musée - atelier textile du feutre de Mouzon." *La lettre de l'OCIM* 58 (1998), pp. 23–25.

Claridge, M. "Photocopiers: An Office Hazard." *Environmental Health* (September 1983), pp. 246–247.

Clarke, S.G., and E.E. Longhurst. "The Corrosion of Metals by Acid Vapours from Wood." *Journal of Applied Chemistry* 11 (1961), pp. 435–443.

Clean Air Network. *Fast Facts on Power Plants*. Washington, DC, 2000.

Conservation by Design Limited, UK. *Danchek Film Indicators*. nd. <www.conservation-by-design.co.uk/danchek.html>

Cooke, R.U., and G.B. Gibbs. *Crumbling Heritage: Studies of Stone Weathering in Polluted Atmospheres*. Swindon, UK: National Power plc, 1994, pp. 48–49.

Cooler Heads Coalition. *Global Warming: Information Page*.

Crawshaw, G.H. "Floorcoverings: The Role of Wool Carpets in Controlling Indoor Air Pollution." *Textile Institute and Industry* 16 (January 1978), pp. 12–15.

Dahlin, E., J.F. Henriksen, and O. Anda. "Assessment of Environmental Risk Factors in Museums and Archives." *European Cultural Heritage Newsletter on Research* 10 (1997), pp. 94–97.

Daniel, F., F. Flieder, and F. Leclerc. "Étude de l'effet de la pollution sur des papiers déacidifiés." pp. 37–72 in *Les documents graphiques et photographiques. Analyse et conservation*. Travaux du CRCDG. Paris: Archives nationales, La Documentation française, 1991.

Daniel, F., F. Flieder, and F. Leclerc. "Étude de l'effet de la pollution sur des papiers déacidifiés." pp. 53–93 in *Les documents graphiques et photographiques: Analyse et conservation*. Travaux du CRCDG. Paris: Archives nationales, La Documentation française, 1988.

Daniel, V., and S. Maekawa. "The Moisture Capacity of Museum Cases." pp. 453–458 in *Materials Research Society Symposium Proceedings, Volume 267* (edited by P.B. Vandiver, J.R. Druzik, G.S. Wheeler, and I.C. Freestone). Pittsburgh: Materials Research Society, 1992.

Daniel, V.D., and S. Ward. "A Rapid Test for the Detection of Substances which will Tarnish Silver." *Studies in Conservation* 27 (1998), pp. 58–60.

Davies, T.D., B. Ramer, G. Kaspyzok, and A.C. Delany. "Indoor/Outdoor Ozone Concentrations at a Contemporary Art Gallery." *Journal of the Air Pollution Control Association* 31 (1984), pp. 135–137.

De Santis, F., I. Allegrini, P. Di Filippo, and D. Pasella. "Simultaneous Determination of Nitrogen Dioxide and Peroxyacetyl Nitrate in Ambient Atmosphere by Carbon-coated Annular Diffusion Denuder." *Atmospheric Environment* 30 (1996), pp. 2637–2646.

Dingle, P., P. Olden, S. Hu, and F. Murray. "A Study of Formaldehyde in a New Office Building." pp. 51–55 in *Proceedings of the 6th International Conference on Indoor Air Quality and Climate, Indoor Air '93, Volume 2* (edited by P. Kalliokoski, J. Jantunen, and O. Seppanen). Helsinki: International Conference on Indoor Air Quality and Climate, 1993.

Donovan, P.D., and T.M. Moynehan. "The Corrosion of Metal by Vapours from Air-drying Paints." *Corrosion Science* 5 (1965), pp. 803–814.

Donovan, P.D., and J. Stringer. "The Corrosion of Metals by Organic Acid Vapours." pp. 537–543 in *Proceedings of the 4th International Congress on Metallic Corrosion* (edited by N.E. Hammer). Houston, TX: National Association of Corrosion Engineers, 1972.

Down, J.L., M.A. MacDonald, J. Tétreault, and R.S. Williams. "Adhesive Testing at the Canadian Conservation Institute — An Evaluation of Selected Poly(Vinyl Acetate) and Acrylic Adhesives." *Studies in Conservation* 41 (1996), pp. 19–44.

Drisko, K., G.R. Cass, P.M. Whitmore, and J.R. Druzik. "Fading of Artists' Pigments due to Atmospheric Ozone." pp. 66–89 in *Wiener Berichte üüber Naturwissenschaft in der Kunst* (edited by A. Vendl, B. Pichler, J. Weber, and G. Banik). Vienna: Verlag ORAC, 1985/86.

Druzik, J. Communication on ozone generators, conservation distribution list. Received March 29, 1999. <palimpsest.stanford.edu/byform/ mailing-lists/cdl/1999/0425.html>

Druzik, J.R., M.S. Adams, C. Tiller, and G.R. Cass. "The Measurement and Model Predictions of Indoor Ozone Concentrations in Museums." *Atmospheric Environment* 24A (1990), pp. 1813–1823.

Duncan, S.J., and V.D. Daniels. *Studies on the Deterioration of Museum Objects by the Release of Formaldehyde from Storage and Display Materials.* Internal report. London: British Museum Department of Conservation, 1986.

Dupont, A.-L., and J. Tétreault. "Exposure of Cellulose to Carboxylic Acid Compounds emitted from Acidic Box." Unpublished data. Ottawa: Canadian Conservation Institute, 2000a.

Dupont, A.-L., and J. Tétreault. "Study of Cellulose Degradation in Acetic Acid Environments." *Studies in Conservation* 45 (2000b), pp. 201–210.

EC (Environment Canada). *Urban Air Quality.* SOE Bulletin No. 99-1. Spring 1999. <www.ec.gc.ca/soer-ree/English/Indicators/ Pdf_file/99-1_e.pdf>

EC (Environment Canada). *National Ambient Air Quality Standards: Proposed Decisions on Particulate Matter and Ozone.* 1997. <www.ec.gc.ca/air/qual/matter.html>

EC (Environment Canada). *Climate Trends and Variations Bulletin.* nd. <www.msc-smc.ec.gc.ca/ccrm/bulletin/>

EC (Environment Canada) and HC (Health Canada). *National Ambient Air Quality Objectives for Ground-level Ozone: Science Assessment Document.* Hull: Environment Canada, 1999, pp. 7-1–7-32. <www.hc-sc.gc.ca/hecs-sesc/air_quality/publications/ ground_level_ozone/part1/toc.htm>

Edwards, C.J., F. Lyth Hudson, and J.A. Hockey. "Sorption of Sulphur Dioxide by Paper." *Journal of Applied Chemistry* 18 (1968), pp. 146–148.

EEA (European Environment Agency). *Corinair 1990.* Last update November 2001.

EMEP (Co-operative Program for Monitoring and Evaluation of the Long-range Transmission of Air Pollutants in Europe). *Emission Maps and Trends for Each Country.* nd. <reports.eea.eu.int/technical_report_2002_91/ en/tab_abstract_RLR>

Environmental Sensors Co. *Formaldehyde Z-300.* nd. <www.environmentalsensors.com/Environmental_ Sensors/Z_Models/Z_Formaldehyde.html>

EPA (Environmental Protection Agency, USA). *Ozone Generators that are sold as Air Cleaners: An Assessment of Effectiveness and Health Consequences.* 2001a. <www.epa.gov/iaq/pubs/ozonegen.html>

EPA. *Latest Findings on National Air Quality: 2000 Status and Trends.* EPA-454/K-01-002. Research Triangle Park, 2001b, pp. 6–13. <www.epa.gov/oar/aqtrnd00/brochure/ 00brochure.pdf>

EPA. *National Air Quality and Emissions Trends Report, 1999.* EPA-454/R-01-004. Research Triangle Park, 2001c, pp. 24–50. <www.epa.gov/oar/aqtrnd99/>

EPA. *Candles and Incense as Potential Sources of Indoor Air Pollution: Market Analysis and Literature Review.* EPA-600/R-01-001. Research Triangle Park, 2001d. <www.epa.gov/ordntrnt/ORD/NRMRL/ Publications/600R01001.htm>

EPA. *National Air Pollutant Emission Trends, 1900-1998.* EPA-454/R-00-002. Research Triangle Park, 2000, pp. ES-3, ES-5, 2.3, 2.12, 3-27, 3-29.

EPA. *Guidance for Using Continuous Monitors in* $PM_{2.5}$ *Monitoring Networks.* EPA-454/R-98-012. Research Triangle Park, 1998. <www.epa.gov/ttn/amtic/files/ambient/ pm25/r-98-012.pdf>

EPA. *National Air Pollutant Emission Trends, 1900-1995.* EPA-454/R-96-007. Research Triangle Park, 1996, p. 2. Executive summary available. <www.epa.gov/oar/emtrnd95/execsumm.html>

EPA. *Figure 1 Top Ten Source Sub-sectors for Europe, 1990.* <reports.eea.eu.int/92-9167-036-7/en/page005.html>

Eremin, K., S. Adams, and J. Tate. "Monitoring of Deposited Particles Levels within the Museum of Scotland: During and After Construction." *The Conservator* 24 (2000), pp. 15–23.

Eremin, K., and P. Wilthew. "Monitoring Concentrations of Organic Gases within the National Museums of Scotland." *SSCR Journal* 9 (1998), pp. 15–19.

Erhardt, D., and M. Mecklemburg. "Relative Humidity Re-examined." pp. 32–38 in *Preventive Conservation: Practice, Theory and Research* (edited by A. Roy and P. Smith). London: International Institute for Conservation of Historic and Artistic Works, 1994.

Erickson, H. *Report on NO_x at the HRHRC.* 1990. <www.ph.utexas.edu/~erickson/nox@hrc.html>

Eriksson, P., L.-G. Johansson, and H. Strandberg. "Initial Stages of Copper Corrosion in Humid Air Containing SO_2 and NO_2." *Journal of the Electrochemical Society* 140 (1993), pp. 53–59.

Feldman, L.H. "Discoloration of Black and White Photographic Prints." *Journal of Applied Photographic Engineering* 7 (1981), pp. 1–9.

Feller, R.L. *Accelerated Ageing: Photochemical and Thermal Aspects.* Research in Conservation No. 4. Marina del Rey: The Getty Conservation Institute, 1994. <www.getty.edu/conservation/resources/20%reports.html>

Ferek, R.J., and P.A. Covert. "Intercomparison of Measurement of Sulfur Dioxide in Ambient Air by Carbonate-impregnated Filters and Teco Pulsed-fluorescence Analyzers." *Journal of Geophysical Research* 102 (1997), pp. 16267–16272.

Fiala, J., L. Cernikovsky, L. Kozakovic, J. Stedman, and K. Stevenson. *Air Quality in the Phare Countries 1997.* Copenhagen: European Environment Agency, 2002, pp. 19–23. <reports.eea.eu.int/topic_report_2001_16/en>

Fiaud, C., and J. Guinement. "The Effect of Nitrogen Dioxide and Chlorine on the Tarnishing of Copper and Silver in the Presence of Hydrogen Sulfide." *Proceedings of the Electrochemical Society* 86 (1986), pp. 280–304.

Franey, J. "Degradation of Copper and Copper Alloys by Atmospheric Sulfur." pp. 306–315 in *Degradation of Metals in the Atmosphere, ASTM STP 965* (edited by S.W. Dean and T.S. Lee). Philadelphia: American Society for Testing and Materials, 1988.

Franey, J.P., T.E. Graedel, and G.W. Kammlott. "The Sulfiding of Copper by Trace Amounts of Hydrogen Sulfide." pp. 383–392 in *Proceedings of International Symposium on Atmospheric Corrosion* (edited by W.H. Ailor). New York: Wiley-Interscience Publication, 1980.

Franey, J.P., G.W. Kammlott, and T.E. Graedel. "The Corrosion of Silver by Atmosphere Sulfurous Gases." *Corrosion Science* 25 (1985), pp. 133–143.

Gibson, L.T., and A.W. Brokerhof. "A Passive Tube-type Sampler for the Determination of Formaldehyde Vapours in Museum Enclosures." *Studies in Conservation* 46 (2001), pp. 289–303.

Gibson, L.T., B.G. Cooksey, D. Littlejohn, and N.H. Tennent. "A Diffusion Tube Sampler for the Determination of Acetic Acid and Formic Acid Vapours in Museum Cabinets." *Analytica Chimica Acta* 341 (1997a), pp. 11–19.

Gibson, L.T., B.G. Cooksey, D. Littlejohn, and N.H. Tennent. "Determination of Acetic Acid and Formic Acid Vapour Concentrations in the Museum Environment by Passive Sampling." *European Cultural Heritage Newsletter on Research* 10 (1997b), pp. 108–112.

Graedel, T.E. "Copper Patinas Formed in the Atmosphere: II. A Qualitative Assessment of Mechanisms." *Corrosion Science* 27 (1987a), pp. 721–740.

Graedel, T.E. "Copper Patinas Formed in the Atmosphere: III. A Semi-quantitative Assessment of Rates and Constraints in the Greater New York Metropolitan Area." *Corrosion Science* 27 (1987b), pp. 741–769.

Graedel, T.E. "Concentrations and Metal Interactions of Atmospheric Trace Gases Involved in Corrosion." pp. 396–401 in *Proceedings of the Ninth International Congress on Metallic Corrosion.* International Congress on Metallic Corrosion, 1984.

Graedel, T.E., J.P. Franey, G.J. Gualtieri, G.W. Kammlott, and D.L. Malim. "On the Mechanism of Silver and Copper Sulfidation by Atmospheric Hydrogen Sulfide and Carbonyl Sulfide." *Corrosion Science* 25 (1985), pp. 1163–1180.

Graedel, T.E., J.P. Franey, and G.W. Kammlott. "Ozone- and Photon-enhanced Atmospheric Sulfidation of Copper." *Science* 224 (1984), pp. 599–601.

Graedel, T.E., G.W. Kammlott, and J.P. Franey. "Carbonyl Sulfide: Potential Agent of Atmospheric Sulfur Corrosion." *Science* 212 (1981), pp. 663–665.

Green, L.R. *Investigation of Sulphide Corrosion of Bronze: Stage II: Source of Hydrogen Sulphide. Report.* 1992/14. London: The British Museum, 1992.

Green, L.R., and D. Thickett. "Testing Materials for Use in the Storage and Display of Antiquities: A Review Methodology." *Studies in Conservation* 40 (1995), pp. 145–152.

Griffith, D.W.T., and B. Galle. "Flux Measurement of NH_3, N_2O and CO_2 Using Dual Beam FTIR Spectroscopy and the Flux-gradient Technique." *Atmospheric Environment* 34 (2000), pp. 1087–1096.

Grosjean, D. "Ambient Levels of Formaldehyde, Acetaldehyde, and Formic Acid in Southern California: Results of a One-year Base-line Study." *Environmental Science & Technology* 25 (1991), pp. 710–715.

Grosjean, D. "Aldehydes, Carboxylic Acids and Inorganic Nitrate during NSMC." *Atmospheric Environment* 22 (1988), pp. 1637–1648.

Grosjean, D., E. Grosjean, and E.L. Williams II."Fading of Artists' Colorants by a Mixture of Photochemical Oxidants." *Atmospheric Environment* 27A (1993), pp. 765–772.

Grosjean, D., P.M. Whitmore, C.P. De Moor, G.R. Cass, and J.R. Druzik. "Fading of Alizarin, and Related Artists' Pigments by Atmospheric Ozone: Reaction Products and Mechanisms." *Environmental Science & Technology* 21 (1987), pp. 635–643.

Grzywacz, C.M., and N.H. Tennent. "Pollution Monitoring in Storage and Display Cabinets: Carbonyl Pollutant Levels in Relation to Artifact Deterioration." pp. 164–170 in *Preventive Conservation: Practice, Theory and Research* (edited by A. Roy and P. Smith). London: International Institute for Conservation of Historic and Artistic Works, 1994.

Harrison, R.M. "Chemistry of the Troposphere." pp. 169–193 in *Pollution: Causes, Effects and Control, third edition* (edited by R.M. Harrison). Cambridge: The Royal Society of Chemistry, 1996.

Havermans, J. "Effects of SO_2 and NO_X on the Accelerated Ageing of Paper." *European Cultural Heritage Newsletter on Research* 10 (1997), pp. 128–133.

Hewings, J. *Air Quality Indices: A Review.* Pollution Probe, 2001, pp. 15–27. <www.pollutionprobe.org/Reports/AQIReport.pdf>

Hisham, M.W.M., and D. Grosjean. "Air Pollution in Southern California Museums: Indoor and Outdoor Levels of Nitrogen Dioxide, Peroxyacetyl Nitrate, Nitric Acid and Chlorinated Hydrocarbons." *Environmental Science & Technology* 25 (1991a), pp. 857–862.

Hisham, M.W.M., and D. Grosjean. "Sulfur Dioxide, Hydrogen Sulfide, Total Reduced Sulfur, Chlorinated Hydrocarbons and Photochemical Oxidant in Southern California Museums." *Atmospheric Environment* 25A (1991b), pp. 1497–1505.

Hjellbrekke, A.-G. *Ozone Measurements 1998.* EMEP/CCC-Report 5/2000. EMEP Co-operative Programme for Monitoring and Evaluation of the Long-Range Transmission of Air Pollutants in Europe, 2000, pp. 17, 53–69. <www.nilu.no/projects/ccc/reports/>

Hoevel, C.L. "A Study of the Discoloration Products Found in White Lead Paint Films." pp. 35–42 in *The 1985 Book and Paper Group Annual* (edited by J. Abt). Washington, DC: The Book and Paper Group, 1985.

Hollinshead, P.W., M.D. Wan Ert, S.C. Holland, and K. Velo. *Deteriorating Negatives: A Health Hazard in Collection Management.* Arizona State Museum, 1987. <www.statemuseum.arizona.edu/coll/phot_neg_hazard.doc>

Howarth, K. "Handling and Care of Modern Audio Tape." pp. 172–176 in *Proceedings of Care of Photographic Moving Image & Sound Collections* (edited by S. Clark). York: Institute of Paper Conservation, 1998.

Image Permanence Institute. *User's Guide for A-D Strips: Film Base Deterioration Monitors.* Rochester: Rochester Institute of Technology, 1997.

ISA (Instrument Society of America). *Environmental Conditions for Process Measurement and Control Systems: Airborne Contaminants.* ISA Standard S71.04-1985. Research Triangle Park: International Society for Measurement and Control, 1986.

Iversen, T., and J. Kolar. *Kvavedioxids effekter pa papper (Effect of Nitrogen Dioxide on Paper)*, FoU-Projektet for papperskonservering, Report 5. Stockholm: Riksarkivet, 1991, pp. 9–27.

Jaffe, L.S. "The Effects of Photochemical Oxidants on Materials." *Journal of the Air Pollution Control Association* 17 (1967), pp. 375–378.

Jarnstrom, H., and K. Saarela. "Indoor Air Quality and Material Emissions in New Buildings." pp. 201–206 in *Proceedings of the 9th International Conference on Indoor Air Quality and Climate, Volume 2* (edited by H. Levin). Santa Cruz: International Conference on Indoor Air Quality and Climate, 2002.

Jensen, B., P. Wolkoff, C.K. Wilkins, and P.A. Clausen. "Characterization of Linoleum. Part I : Measurement of Volatile Organic Compounds by Use of the Field and Laboratory Emission Cell, FLEC." pp. 443–448 in *Proceedings of the 6th International Conference on Indoor Air Quality and Climate, Indoor Air '93, Volume 2* (edited by P. Kalliokoski, J. Jantunen, and O. Seppanen). Helsinki: International Conference on Indoor Air Quality and Climate, 1993.

Johansson, E., C. Leygraf, and B. Rendahl. "Characterisation of Corrosivity in Indoor Atmospheres with Different Metals and Evaluation Techniques." *British Corrosion Journal* 33 (1998), pp. 59–66.

Johansson, E., B. Rendahl, V. Kucera, C. Leygraf, D. Knotkova, J. Vlckova, and J. Henriksen. "Comparison of Different Methods for Assessment of Corrosivity in Indoor Environments." *European Cultural Heritage Newsletter on Research* 10 (1997), pp. 92–94.

Johansson, L.-G., O. Lindqvist, and R.E. Mangio. "Corrosion of Calcareous Stones in Humid Air Containing SO_2 and NO_2." *Durability of Building Materials* 5 (1988), pp. 439–449.

Jones, N.C., C.A. Thornton, D. Mark, and R.M. Harrison. "Indoor/Outdoor Relationships of Particulate Matter in Domestic Homes with Roadside, Urban and Rural Locations." *Atmospheric Environment* 34 (2000), pp. 2603–2612.

Judeiki, H.S., and T.B. Stewart. "Laboratory Measurement of SO_2 Deposition Velocities on Selected Building Materials and Soils." *Atmospheric Environment* 10 (1976), pp. 769–776.

K & M ChromAir. *ChromAir Direct Read Passive Monitoring Badges.* nd. <www.kandmenvironmental.com/Products.html>

Kadokura, T., K. Yoshizumi, M. Kashiwagi, and M. Saito. "Concentration of Nitrogen Dioxide in the Museum Environment and its Effects on the Fading of Dyed Fabrics." pp. 87–89 in *The Conservation of Far Eastern Art, Preprints of the Contributions to the Kyoto Congress* (edited by J.S. Mills, P. Smith, and K. Yamasaki). London: International Institute for Conservation of Historic and Artistic Works, 1988.

Kames, J. McLeod Russel Filter AG, Switzerland. Personal communication, 2002.

Kames, J. "Filtration Systems in Cultural and Heritage Collections." In *Indoor Air Quality and Climate in Cultural and Heritage Institutions.* Workshop 21, Espoo, Finland, Healthy Buildings 2000, 2000. <www.hb2000.org/workshop21.html>

Keepsafe Systems. *Case Leakage Tester.* 2002a. <www.microclimate.ca/product%20data/caseleakage.htm>

Keepsafe Systems. *Micro Climate Technology.* 2002b. <www.microclimate.ca/product%20data/mcg%20overview.htm>

Kepner, C.H., and B.B. Tregoe. *The Rational Manager: A Systematic Approach to Problem Solving and Decision Making, second edition.* New Jersey: Kepner-Tregoe, Inc., 1976, pp. 174–206.

Kildeso, J., J. Vallarino, J.D. Spengler, H.S. Brightman, and T. Schneider. "Dust Build-up on Surfaces in the Indoor Environment." *Atmospheric Environment* 33 (1999), pp. 699–707.

Kirchner, S., A.M. Laurent, B. Collignan, Y. Le Moullec, O. Ramalho, J.G. Villenave, and J.P. Flori. "Impact of the Urban Pollution on the Indoor Environment — Experimental Study on a Mechanical Ventilation Dwelling." pp. 164–169 in *Proceedings of the 9th International Conference on Indoor Air Quality and Climate, Volume 1* (edited by H. Levin). Santa Cruz: International Conference on Indoor Air Quality and Climate, 2002.

Kleno, J.G., P.A. Clausen, C.J. Weschler, and P. Wolkoff. "Determination of Ozone Removal Rates by Selected Building Products using the FLEC Emission Cell." *Environmental Science & Technology* 35 (2001), pp. 2548–2553.

Kontozova, V., F. Deutsch, R. Godoi, A.F. Godoi, P. Joos, and R. Van Grieken. "Characterisation of Air Pollutants in Museum Showcases." In *Proceedings of the 7th International Conference on Non-destructive Testing and Microanalysis for the Diagnostics and Conservation of the Cultural and Environmental Heritage* (Antwerp, Belgium, June 2–6, 2002). <www.uia.ac.be/u/vgrieken/pages/Kontozovacorr.pdf>

Kozdron-Zabiegala, B., J. Namiesnik, and A. Przyjazny. "Use of Passive Dosimeters for Evaluation of the Quality of Indoor and Outdoor Air." *Indoor Environment* 4 (1995), pp. 189–203.

Koziel, J.A., J. Noah, and J. Pawliszyn. "Field Sampling and Determination of Formaldehyde in Indoor Air with Solid-phase Microextraction and On-fiber Derivatization." *Environmental Science & Technology* 35 (2001), pp. 1481–1486.

Kruse, K., L. Madso, A. Skogstad, W. Eduard, F. Levy, K. Skulberg, and K. Skyberg. "Characterization of Particle Types in Modern Offices." pp. 880–884 in *Proceedings of the 9th International Conference on Indoor Air Quality and Climate, Volume 1* (edited by H. Levin). Santa Cruz: International Conference on Indoor Air Quality and Climate, 2002.

Kuhn, U., S. Rottenberger, T. Biesenthal, C. Ammann, A. Wolf, G. Schebeske, A. Gut, F.X. Meixer, P.S.T. Oliva, T.M. Tavares, and J. Kesselmeier. "Exchange of Short-chained Monocarboxylic Acids between Biosphere and the Atmosphere at a Remote Tropical Forest in Amazonia." In *Volatile Organic Compounds (VOC) produced by Plants*. PhytoVOC 2001, 2001. <phytovoc.cefe.cnrs-mop.fr/abstracts.htm>

Kukadia, V., J. Palmer, J. Littler, R. Woolliscroft, R. Watkins, and I. Ridley. "Air Pollution Levels Inside Buildings in Urban Areas: A Pilot Study." pp. 322–332 in *Proceedings of CIBSE/ASHRAE Joint National Conference Part Two, Volume 1*. 1996.

Ladisch, C.M., and S.-L. Rau. "The Effect of Humidity on the Ozone Fading of Acid Dyes." *Textile Chemist and Colorist* 29 (1997), pp. 24–28.

Lafontaine, R.H. *Silica Gel*. Technical Bulletin, No. 10. Ottawa: Canadian Conservation Institute, 1984.

Lal Gauri, K., A.N. Chowdhury, N.P. Kulshreshtha, and A.R. Punuru. "The Sulfations of Marble and the Treatment of Gypsum Crusts." *Studies in Conservation* 34 (1989), pp. 201–206.

Lanting, R.W. "Air Pollution in Archives and Museums: Its Pathway and Control." pp. 665–670 in *Preprints of Indoor Air 90, Volume 3*. Ottawa: International Conference on Indoor Air Quality and Climate, 1990.

Larsen, R. "Deterioration and Conservation of Vegetable Tanned Leather." *European Cultural Heritage Newsletter on Research* 10 (1997), pp. 54–61.

Lee, K., J. Vallarino, T. Dumyahn, H. Ozkaynak, and J.D. Spengler. "Ozone Decay Rates in Residences." *Journal of the Air and Waste Management Association* 49 (1999), pp. 1238–1244.

Leichnitz, K. *Detector Tube Handbook: Air Investigations and Technical Gas Analysis with Dräger Tubes, seventh edition*. Lübeck: Dräger, 1989.

Leissner, J. "Assessment and Monitoring the Environment of Cultural Property." *European Cultural Heritage Newsletter on Research* 10 (1997), pp. 5–49.

Leveque, M.A. "The Problem of Formaldehyde — A Case Study." pp. 56–65 in *Preprints of the Annual Meeting of the AIC* (edited by A.G. Brown). Williamsburg: The American Institute for Conservation of Historic and Artistic Works, 1986.

Leyshon, L.J., and C. Holstead. "Reaction of Sulphur Dioxide with Image Transfer Azo-Naphthol Dyes." *Journal of Photographic Science* 36 (1988), pp. 107–114.

Little, B.J., P.A. Wagner, and Z. Lewandowski. "The Role of Biomineralization in Microbiologically Influenced Corrosion." pp. 294/1–294/18 in *Corrosion 98*. NACE International, 1998.

Lobnig, R.E., R.P. Frankenthal, D.J. Siconolfi, and J.D. Sinclair. "The Effect of Submicron Ammonium Sulfate Particles on the Corrosion of Copper." pp. 336–356 in *Corrosion and Reliability of Electronic Materials and Devices*. Pennington: The Electrochemical Society Inc., 1993.

Lobnig, R.E., D.J. Siconolfi, J. Maisano, G. Grundmeier, H. Streckel, R.P. Frankenthal, M. Stratmann, and J.D. Sinclair. "Atmospheric Corrosion of Aluminum in the Presence of Ammonium Sulfate Particles." *Journal of the Electrochemical Society* 143 (1996a), pp. 1175–1181.

Lobnig, R.E., D.J. Siconolfi, L. Psota Kely, G. Grundmeier, R.P. Frankenthal, M. Stratmann, and J.D. Sinclair. "Atmospheric Corrosion of Zinc in the Presence of Ammonium Sulfate Particles." *Journal of the Electrochemical Society* 143 (1996b), pp. 1539–1546.

López-Delgado, A., E. Cano, J.M. Bastidas, and F.A. López. "A Laboratory Study of the Effect of Acetic Acid Vapor on Atmospheric Copper Corrosion." *Journal of the Electrochemical Society* 145 (1998), pp. 4140–4147.

Lorenzen, J.A. "Atmospheric Corrosion of Silver." pp. 110–116 in *Proceedings of the 17th Annual Technical Meeting of the Institute of Environmental Sciences.* Los Angeles: Institute of Environmental Sciences, 1971.

Luke, W.T. "Evaluation of a Commercial Pulsed Fluorescence Detector for the Measurement of Low-level SO_2 Concentrations during the Gas-phase Sulfur Intercomparison Experiment." *Journal of Geophysical Research* 102 (1997), pp. 16255–16265.

Lynn, Y., G. Salmon, and G.R. Cass. "The Ozone Fading of Traditional Chinese Plant Dyes." *Journal of AIC* 39 (2000), pp. 245–257.

Lyth Hudson, F. "Acidity of 17th and 18th Century Books in Two Libraries." *Paper Technology* 8 (1967), pp. 189, 190, and 196.

Maekawa, S. (ed.). *Oxygen-free Museum Cases.* Research in Conservation. Los Angeles: The Getty Conservation Institute, 1998. <www.getty.edu/conservation/resources/reports.html>

Manahan, S.E. *Environmental Chemistry, sixth edition.* Boca Raton: Lewis Publishers, 1994, p. 390.

Marshall, G.B., and N.A. Dimmock. "Determination of Nitric Acid in Ambient Air using Diffusion Denuder Tubes." *Talanta* 39 (1992), pp. 1463–1469.

Martin, G. "Air Exchange Rates in Display Cases — A Standard?" In *Proceedings of the Third Indoor Air Pollution Meeting* (edited by M. Ryhl-Svendsen). Indoor Air Quality Working Group, 2000. <iaq.dk/iap/iaq2000/2000_04.htm>

Matthews, T.G., T.J. Reed, B.J. Tromberg, C.E. Daffron, and A.R. Hawthorne. "Formaldehyde Emission from Combustion Source and Solid Formaldehyde-resin-containing Products: Potential Impact on Indoor Formaldehyde Concentrations." pp. 131–150 in *Formaldehyde: Analytical Chemistry Toxicology* (edited by V. Turoski). Washington, DC: American Chemical Society, 1985.

Mazurkiewicz, G. *CEA Instruments, Inc. Formaldehyde Monitor.* 2001. <www.achrnews.com/CDA/ArticleInformation/products/BNPProductItem/0,6080,21157,00.html>

Meyer, B., and K. Hermanns. "Formaldehyde Release from Wood Products: An Overview." pp. 1–16 in *Formaldehyde Release from Wood Products, ACS Symposium Series #316* (edited by B. Meyer, B.A. Kottes Andrews, and R.M. Reinhardt). Washington, DC: American Chemical Society, 1986.

Meyer, B., and K. Hermanns. "Formaldehyde Release from Pressed Wood Products." pp. 101–116 in *Formaldehyde: Analytical Chemistry Toxicology* (edited by V. Turoski). Washington, DC: American Chemical Society, 1985.

Michalski, S. *Guidelines for Humidity and Temperature for Canadian Archives.* CCI Technical Bulletin, No. 23. Ottawa: Canadian Conservation Institute, 2000, pp. 6–7.

Michalski, S. *ISO Blue Wool Standard Equivalency.* Ottawa: Canadian Conservation Institute, unpublished data, 1999.

Michalski, S. "Lighting Decision." pp. 97–104 in *Fabric of an Exhibition: An Interdisciplinary Approach - Preprints.* Ottawa: Canadian Conservation Institute, 1997.

Michalski, S. "A Systematic Approach to Preservation: Description and Integration with Other Museum Activities." pp. 8–11 in *Preventive Conservation: Practice, Theory and Research* (edited by A. Roy and P. Smith). London: International Institute for Conservation of Historic and Artistic Works, 1994a.

Michalski, S. "Leakage Prediction for Buildings, Cases, Bags and Bottles." *Studies in Conservation* 39 (1994b), pp. 169–186.

Michalski, S. "Sharing Responsibility for Conservation Decisions." pp. 241–258 in *Durability and Change: The Science, Responsibility, and Cost of Sustaining Cultural Heritage* (edited by W.E. Krumbein, P. Brimblecombe, D.E. Cosgrove, and S. Staniforth). Toronto: John Wiley & Sons, 1994c.

Michalski, S. "Damage to Museum Objects by Visible Radiation (Light) and Ultraviolet Radiation (UV)." pp. 3–16 in *Lighting: A Conference on Lighting in Museums, Galleries, and Historic Houses.* Bristol: The Museums Association, United Kingdom Institute for Conservation and Group of Designers and Interpreters in Museums, 1987.

Michalski, S. "A Control Module for Relative Humidity in Display Cases." pp. 28–31 in *Science and Technology in the Service of Conservation* (edited by N.S. Brommelle, N. Stolow, and G. Thomson). London: International Institute for Conservation of Historic and Artistic Works, 1982.

MIE (Monitoring Instruments for the Environment, Inc.). *DataRAM 4; Portable Particle Sizing Aerosol Monitor/Data Logger.* 2002. <www.thermo.com/cda/product/detail/1,1055,22453,00.html>

Miles, C. "Wood Coatings for Display and Storage Cases." *Studies in Conservation* 31 (1986), pp. 114–126.

Ministère de l'environnement et de la faune (Quebec). *Air Quality in Québec: 1975-1994.* Ministère de l'environnement, 1997, pp. 23–24. <www.menv.gouv.qc.ca/air/qualite-en/index.htm>

MOE (Ministry of the Environment, Ontario). *Air Quality in Ontario.* Ministry of the Environment, 1999, pp. 24–25. <www.ene.gov.on.ca/air.htm>

MOE. *A Compendium of Current Knowledge on Fine Particle Matter in Ontario.* Ministry of the Environment, 1995, pp. III.2–III.14. <www.ene.gov.on.ca/envision/env%5Freg/er/documents/pa9e0008/list.htm>

Molhave, L. "Volatile Organic Compounds, Indoor Air Quality and Health." pp. 15–33 in *Preprints of Indoor Air 90, Volume 5.* Ottawa: International Conference on Indoor Air Quality and Climate, 1990.

Moore, A., F. Ruetsch, and H.-D. Weigmann. "The Role of Dye Diffusion in the Ozone Fading of Acid and Disperse Dyes in Polyamides." *Textile Chemist and Colorist* 16 (1984), pp. 250–256.

Moroni, B., and G. Poli. "Corrosion of Limestone in Humid Air Containing SO_2 and NO_2: Result after Short-term Laboratory Experiments." *Science and Technology for Cultural Heritage* 5 (1996), pp. 7–18.

Muller, C.O. *Achieving Your Indoor Air Quality Goals: Which Filtration System Works Best?* Purafil Inc., nda, p. 6. <www.afslasvegas.com/purafil/catalog2/literature/assets/pdf/pur/achieve.htm>

Muller, C.O. *Gaseous Contamination Control Strategies at The Hague.* Purafil Inc., ndb, pp. 8 and 16.

NAST (National Assessment Synthesis Team). *Climate Change Impacts on the United States: The Potential Consequences of Climate Variability and Change.* New York: Cambridge University Press, 2001, pp. 22 and 447. <www.usgcrp.gov/usgcrp/Library/nationalassessment/foundation.htm>

Nathanson, T., L. Morawska, and M. Jamriska. "Filter Performance Guidelines for Good IAQ." pp. 1082–1087 in *Proceedings of the 9th International Conference on Indoor Air Quality and Climate, Volume 2* (edited by H. Levin). Santa Cruz: International Conference on Indoor Air Quality and Climate, 2002.

Nazaroff, W.W., M.P. Ligocki, L.G. Salmon, G.R. Cass, T. Fall, M.C. Jones, H.I.H. Liu, and T. Ma. *Airborne Particles in Museums.* Research in Conservation No. 6. Marina del Rey: The Getty Conservation Institute, 1993, pp. 38–41 and 91–103. <www.getty.edu/conservation/resources/reports.html>

Nazaroff, W.W., M.P. Ligocki, L.G. Salmon, G.R. Cass, T. Fall, M.C. Jones, H.I.H. Liu, and T. Ma. *Protection of Works of Art from Soiling due to Airborne Particulates.* GCI Scientific Program Report. Marina del Rey: The Getty Conservation Institute, 1992, pp. 9–13.

NewScientist. *Climate Change - Web resources.* <www.newscientist.com/hottopics/climate/climatelinks.jsp>

Newton, L.R., W.H. Anderson, H.S. Lagroon, and K.A. Stephens. "Large-scale Test Chamber Methodology for Urea-formaldehyde Bonded Wood Products." pp. 154–187 in *Formaldehyde Release from Wood Products, ACS Symposium Series #316* (edited by B. Meyer, B.A. Kottes Andrews, and R.M. Reinhardt). Washington, DC: American Chemical Society, 1986.

NFPA (National Fire Protection Association). *Storage and Handling of Cellulose Nitrate Film.* NFPA 40. Quincy, MA: NFPA, 2001.

Nicholson, C., and E. O'Loughlin. *Screening Conservation, Storage and Exhibit Materials using Acid-detection Strips.* Saint Paul, MN: Northern States Conservation Center, 2000. <www.collectioncare.org/pubs/v1n4p4.html>

Nicholson, C., and E. O'Loughlin. "The Use of A-D Strips for Screening Conservation and Exhibit Materials." pp. 83–84 in *The Book and Paper Group Annual* (edited by R. Espinosa). Provo, UT: AIC, 1996.

Nguyen, T.-P., B. Lavédrine, and F. Flieder. "Effets de la pollution atmosphérique sur la dégradation de la gélatine photographique." pp. 567–571 in *Preprints of ICOM-CC Conference, Volume 2* (edited by J. Bridgland). London: James & James Science Publishers Ltd., 1999.

Oakley, V. "Vessel Glass Deterioration at the Victoria and Albert Museum: Surveying the Collection." *The Conservator* 14 (1990), pp. 30–36.

Odlyha, M., O.F. Van den Brink, J.J. Boon, M. Bacci, and N. Cohen. "Damage Assessment of Museum Environments using Paint-based Dosimetry." pp. 73–79 in *Preprints of ICOM-CC Conference, Volume 1* (edited by R. Vontobel). London: James & James Science Publishers Ltd., 2002.

Ordonez, E., and J. Twilley. "Clarifying the Haze: Efflorescence on Works of Art." *WAAC Newsletter* 20 (1998), pp. 12–17.

Organ, R.M. "Remarks on Inhibitors used in Steam Humidification." *Bulletin of the IIC - American Group* 7 (1967), p. 31.

Oshio, R. "Contamination Control of Alkaline Substance in Newly Built Museums." *Scientific Papers on Japanese Antiques and Art Crafts* 37 (1992), pp. 54–59.

Pacaud, G. "Aperçu sur la désinsectisation par anoxie sous atmosphere inerte : 3- les films haute barrière à l'oxygène." *La lettre de l'OCIM* 62 (1999), pp. 23–28.

Pacaud, G. "Aperçu sur la désinsectisation par anoxie sous atmosphere inerte : 2- système dynamique (suite)." *La lettre de l'OCIM* 59 (1998), pp. 29–36.

Padfield, T. "The Control of Relative Humidity and Air Pollution in Show-cases and Picture Frames." *Studies in Conservation* 11 (1966), pp. 9–29.

Padfield, T., M. Burke, and D. Erhardt. "A Cooled Display Case for George Washington's Commission." pp. 38–42 in *ICOM-CC Preprints* (edited by D. de Froment). Paris: ICOM, Committee for Conservation, 1984.

Parmar, S.S., and D. Grosjean. "Sorbent Removal of Air Pollutants from Museum Display Cases." *Environmental International* 17 (1991), pp. 39–50.

Parmar, S.S., and D. Grosjean. *Removal of Air Pollutants from Museum Display Cases: Final Report, August 1989.* Marina del Rey: Getty Conservation Institution, 1989, pp. 3–15.

Pauly, S. "Permeability and Diffusion Data." pp. VI-437–VI-445 in *Polymer Handbook, third edition* (edited by J. Brandrup and E.H. Immergut). New York: John Wiley & Sons, 1989.

Payrissat, M., and S. Beilke. "Laboratory Measurements of the Uptake of Sulfur Dioxide by Different European Soils." *Atmospheric Environment* 9 (1975), pp. 211–217.

Perkins, L. "A Model for Humidity Control using Gore-Tex Silica Tiles." pp. 100–105 in *The Book and Paper Group Annual, Volume 6* (edited by R. Espinosa). Washington, DC: Book and Paper Group of the American Institute for Conservation of Historic and Artistic Works, 1987.

Persson, D., and C. Leygraf. "Metal Carboxylate Formation during Indoor Atmospheric Corrosion of Cu, Zn, and Ni." *Journal of the Electrochemical Society* 142 (1995), pp. 1468–1477.

Phibbs, H. "Microenclosures for Framed Collections." *Exhibitionist* 20 (2001), pp. 37–40.

Pilz, M. "Umweltsituation im Grunen Gewolbe Dresden: Charakterisierung mit Hilfe von Glassensoren." *Restauro* 106 (2000), pp. 422–426.

Pope, D., H.R. Gibbens, and R.L. Moss. "The Tarnishing of Ag at Naturally-occurring H_2S and SO_2 Levels." *Corrosion Science* 8 (1968), pp. 883–887.

Popp, C.J., and R.S. Martin. *Reactive Atmospheric Organic Compounds in the El Paso, Texas-Ciudad Juarez, Mexico Airshed.* SCERP project number: AQ94-23. San Diego: Southwest Center for Environmental Research and Policy, 1999. <www.scerp.org/scerp/projects/AQ94_2.3.html>

Purafil. *Media Technical Brochure.* Doraville: Purafil, 1999, p. 8. <www.afslasvegas.com/purafil/catalog2/literature/assets/pdf/pur/mediatec.htm>

Purafil. *OnGuard Atmospheric Corrosion Monitor.* Doraville: Purafil, 1998a. <www.afslasvegas.com/purafil/catalog2/literature/assets/pdf/pur/onguard1.htm>

Purafil. *Purafil's Technical Services Support.* Doraville: Purafil, 1998b. <www.afslasvegas.com/purafil/catalog2/literature/assets/pdf/pur/techsrv1.htm>

Raphael, T., N. Davis, and K. Brookes. *Exhibit Conservation Guidelines: Incorporating Conservation into Exhibit Planning, Design and Fabrication.* CD-ROM. U.S. National Park Service, 1999, pp. C:3, C:4, 3:7, 5:A, 5:H.

Reilly, J. *IPI Storage Guide for Acetate Film: Instructions for using the Wheel, Graphs, and Table: Basic Strategy for Film Preservation.* Rochester: Rochester Institute of Technology, Image Permanence Institute, 1993.

Reilly, J.M. *Storage Guide for Color Photographic Materials: Caring for Color Slides, Prints, Negatives, and Movie Films.* Albany: The University of the State of New York, 1998, pp. 12 and 19–23.

Reilly, J.M., D.W. Nishimura, and E. Xinn. *New Tools for Preservation: Assessing Long-term Environmental Effects on Library and Archives Collections.* Washington, DC: The Commission on Preservation and Access, 1995, pp. 7, 21–29.

Reilly, J.M., E. Zinn, and P. Adelstein. *Atmospheric Pollutant Aging Test Method Development.* Final report to American Society for Testing and Materials. Rochester: Image Permanence Institute, 2001, pp. 52–97.

Reiner, T., O. Mohler, and F. Arnold. "Measurement of Acetone, Acetic Acid, and Formic Acid in the Midlatitude Upper Troposphere and Lower Stratosphere." *JGR-Atmosphere* 104 (1999), pp. 139–143.

Riederer, J. "Environmental Damage to Museum Objects." *European Cultural Heritage Newsletter on Research* 10 (1997), pp. 118–121.

Rodgers, J. "Preservation and Conservation of Video Tape." pp. 6–10 in *Preprints of Care of Photographic Moving Image & Sound Collections* (edited by S. Clark). York: Institute of Paper Conservation, 1998.

Roemich, H. Personal communication, 2002. Glass sensors available at the Fraunhofer-Institut für Silicatforschung, Wertheim-Bronnbach, Germany.

Rogers, G. De W., and C.G. Costain. "Contamination of a Collection by a Demineralizer/Aerosol Humidifier System." *Journal of IIC-GC* 5 (1980), pp. 22–24.

Ryan, J., D. McPhail, P. Rogers, and V. Oakley. "Glass Deterioration in the Museum Environment." *Chemistry and Industry* 13 (1993), pp. 498–501.

Ryhl-Svendsen, M. *Concentration Converters.* IAQ in Museums and Archives. 2001. <iaq.dk/papers/conc_calc.htm>

Ryhl-Svendsen, M. *Indoor Air Pollution in Museums: Detection of Formic and Acetic Acid Emission from Construction Materials by SPME-GC/MS.* Master's thesis. Copenhagen: School of Conservation, The Royal Danish Academy of Fine Arts, 2000, pp. 66–86.

Ryhl-Svendsen, M., and J. Glastrup. "Acetic and Formic Acid Concentrations in the Museum Environment measured by SPME-GC/MS." *Atmospheric Environment* 36 (2002), pp. 3909–3916.

Saaty, T.L. *The Analytic Hierarchy Process.* Pittsburgh: RWS Publications, 1990, pp. 1–35.

Sabersky, R.H., D.A. Sinema, and F.H. Shair. "Concentrations, Decay Rates, and Removal of Ozone and their Relation to establishing Clean Indoor Air." *Environmental Science & Technology* 7 (1973), pp. 347–353.

Saito, M., S. Goto, and M. Kashiwagi. "Effect of the Concentration of NO₂ Gas to the Fading of Fabrics dyed with Natural Dyes." *Scientific Papers on Japanese Antiques and Art Crafts* 39 (1994), pp. 67–74.

Saito, M., S. Goto, and M. Kashiwagi. "Effect of the Composition of NO₂ Gas to the Fading of Plant Dyes." *Scientific Papers on Japanese Antiques and Art Crafts* 38 (1993), pp. 1–9.

Salmon, L.G., and G.R. Cass. "The Fading of Artists' Colorants by Exposure to Atmospheric Nitric Acid." *Studies in Conservation* 38 (1993), pp. 73–91.

Salmon, L.G., W.W. Nazaroff, M.P. Ligocki, M.C. Jones, and G.R. Cass. "Nitric Acid Concentrations in Southern California Museums." *Environmental Science & Technology* 24 (1990), pp. 1004–1013.

Sano, C. "Indoor Air Quality in Museums: Their Existing Levels, Desirable Conditions and Countermeasures." *Journal of Japan Air Cleaning Association* 38 (2000), pp. 20–26. [In Japanese.]

Saunders, D. "The Environment and Lighting in the Sainsbury Wing of the National Gallery." pp. 630–635 in *Preprints of ICOM-CC Conference, Volume 2* (edited by J. Bridgland). London: James & James Science Publishers Ltd., 1993.

Saunders, D., and J. Kirby. "Light-induced Damage: Investigating the Reciprocity Principle." pp. 87–90 in *Preprints of ICOM-CC Conference, Volume 1* (edited by J. Bridgland). London: James & James Science Publishers Ltd., 1996.

Saunders, D., and J. Kirby. "Light-induced Colour Changes in Red and Yellow Lake Pigments." *National Gallery Technical Bulletin* 15 (1994), pp. 79–97.

Schmith Etkin, D. *Office Furnishings/Equipment & IAQ: Health Impacts, Prevention & Mitigation.* Arlington, MA: Cutter Information Corp., 1992, pp. 89–91.

Schubert, R. "A Second Generation Accelerated Atmospheric Corrosion Chamber." pp. 374–384 in *Degradation of Metals in the Atmosphere, ASTM STP 965* (edited by S.W. Dean and T.S. Lee). Philadelphia: American Society for Testing and Materials, 1988.

Schubert, R., and S.M. D'Egidio. "The Surface Composition of Copper with Indoor Exposures ranging from 3 to 49 Years." *Corrosion Science* 30 (1990), pp. 999–1008.

Seabright, L.H., and J. Trezek. "Notes on the Prevention of White Powder Corrosion of Cadmium Plate." *Plating* (1948), pp. 715–718.

Seifert, B. "Regulating Indoor Air." pp. 35–49 in *Preprints of Indoor Air 90, Volume 5.* Ottawa: International Conference on Indoor Air Quality and Climate, 1990.

Seinfeld, J.H. *Atmospheric Chemistry and Physics of Air Pollution.* Toronto: John Wiley & Sons, 1986, p. 37.

Seinfeld, J.H., and S.N. Pandis. *Atmospheric Chemistry and Physics: From Air Pollution to Climate Change.* Toronto: John Wiley & Sons, 1998, pp. 523–531.

Selwitz, C. *Cellulose Nitrate in Conservation.* Research in Conservation No. 2. Marina del Rey: The Getty Conservation Institute, 1988. <www.getty.edu/conservation/resources/reports.html>

Selwitz, C., and S. Maekawa. *Inert Gases in the Control of Museum Insect Pests.* Research in Conservation. Marina del Rey: The Getty Conservation Institute, 1998. <www.getty.edu/conservation/resources/reports.html>

Shashoua, Y. "Ageless Oxygen Absorber: From Theory to Practice." pp. 881–887 in *Preprints of ICOM-CC Conference, Volume 1* (edited by J. Bridgland). London: James & James Science Publishers Ltd., 1999.

Shaver, C.L., and G.R. Cass. "Ozone and the Deterioration of Works of Art." *Environmental Science & Technology* 17 (1983), pp. 748–752.

Simon, D., J. Bardole, and M. Bujor. "Study of the Reactivity of Silver, Copper, Silver-Copper, and Silver Palladium Alloys used in Telephone Relay Contacts." *IEEE Transactions on Components, Hybrids, and Manufacturing Technology* CHMT-3 (1980), pp. 13–16.

Slinn, W.G., N. Hasse, L. Hicks, B.B.A.W. Hogan, D. Lai, P.S. Liss, K.O. Munnich, G.A. Sehmel, and O. Vittori. "Some Aspects of the Transfer of Atmospheric Trace Constituents past the Air–Sea Interface." *Atmospheric Environment* 12 (1978), pp. 2055–2087.

Smith, L.E. "Factors Governing the Long-term Stability of Polyester-based Recording Media." *Restaurator* 12 (1991), pp. 201–218.

Société Suisse des ingénieurs en chauffage et climatisation. *Directive 96-4 pour l'utilisation des filtres dans les installations aérotechniques.* Bern: Société Suisse des ingénieurs en chauffage et climatisation, 1998.

Souza, S.R., P.C. Vasconcellos, and L.R.F. Carvalho. "Low Molecular Weight Carboxylic Acids in an Urban Atmosphere: Winter Measurements in Sao Paulo City, Brazil." *Atmospheric Environment* 33 (1999), pp. 2563–2574.

Spedding, D.J., and R.P. Rowlands. "Sorption of Sulphur Dioxide by Indoor Surfaces -1: Wallpapers." *Journal of Applied Chemistry* 20 (1970), pp. 143–146.

Stadler-Salt, N., and P. Bertram. "State of the Lakes Ecosystem Conference 2000: Implementing Indicators." Draft for review. Burlington, Chicago: Environment Canada, U.S. Environmental Protection Agency, 2000, p. 89. <www.on.ec.gc.ca/solec/implementing2000-e.html>

Stavroudis, C. "HEPA HEPA HEPA." *AIC Newsletter* January (2002a), pp. 24–25. <aic.stanford.edu/health/hepa.html>

Stavroudis, C. "Never Mind the Bollocks, Here's the HEPA Chart." *WAAC Newsletter* 24 (2002b), pp. 13–16.

Stigbrand, M., A. Karlsson, and K. Irgum. "Direct and Selective Determination of Atmospheric Gaseous Hydrogen Peroxide by Diffusion Scubber and 1,1'-Oxalyldiimidazole Chemiluminescence Detection." *Analytical Chemistry* 68 (1996), pp. 3945–3950.

Tennent, N.H., and T. Baird. "The Identification of Acetate Efflorescence on Bronze Antiquities stored in Wooden Cabinets." *The Conservator* 16 (1992), pp. 39–43 and 47.

Tétreault, J. "Guidelines for Selecting and Using Coatings." *CCI Newsletter* 28 (2001), pp. 5–7.

Tétreault, J. *Coatings for Display and Storage in Museums.* CCI Technical Bulletin, No. 21. Ottawa: Canadian Conservation Institute, 1999a.

Tétreault, J. *Oak Display Cases: Conservation Problems and Solutions.* Special CCI Note. Ottawa: Canadian Conservation Institute, 1999b, 10 pp. <www.cci-icc.gc.ca/document-manager/view-document_e.cfm?Document_ID=80&ref=co>

Tétreault, J. "Corrosion of Zinc and Copper by Acetic Acid Vapor at 54% RH." Unpublished results. Ottawa: Canadian Conservation Institute, 1992a.

Tétreault, J. "La mesure de l'acidité des produits volatils." *Journal of IIC-CG* 17 (1992b), pp. 17–25. <www.cci-icc.gc.ca/document-manager/view-document_e.cfm?Document_ID=81&ref=co>

Tétreault, J. "VOC Emission from Silicone Sealants GE II." Unpublished results. Ottawa: Canadian Conservation Institute, 1990.

Tétreault, J., E. Cano, M. Van Bommel, D. Scott, M.-G. Barthés-Labrousse, L. Minel, and L. Robbiola. "Corrosion of Copper and Lead by Formaldehyde, Formic and Acetic Acid Vapours." *Studies in Conservation* 48 (2003), pp. 231–250.

Tétreault, J., and W.F. Lai. "Discoloration of Colorants by Sulfur Dioxide and Acetic Acid." Unpublished results. Ottawa: Canadian Conservation Institute, 2001.

Tétreault, J., J. Sirois, and E. Stamatopoulou. "Study of Lead Corrosion in Acetic Acid Environment." *Studies in Conservation* 43 (1998), pp. 17–32.

Tétreault, J., and E. Stamatopoulou. "Determination of Concentrations of Acetic Acid emitted from Wood Coatings in Enclosures." *Studies in Conservation* 42 (1997), pp. 141–156.

Tétreault, J., and R.S. Williams. "Crystals on Cellulose Nitrate." Unpublished results. Ottawa: Canadian Conservation Institute, 2002.

Thickett, D. The British Museum. Personal communication, 2000.

Thickett, D. *Relative Effects of Formaldehyde, Formic and Acetic Acids on Lead, Copper and Silver.* The British Museum, Report 1997/12. London: The British Museum, 1997.

Thickett, D., S. Bradley, and L. Lee. "Assessment of the Risks to Metal Artifacts posed by Volatile Carbonyl Pollutants." pp. 260–264 in *Proceedings of the International Conference on Metals Conservation* (edited by W. Mourey and L. Robbiola). London: James & James Science Publishers Ltd., 1998.

Thomson, G. *The Museum Environment, second edition.* London: Butterworths, 1986, pp. 232–236 and 257–258.

Thomson, G. "Stabilization of RH in Exhibition Cases: Hygrometric Half-time." *Studies in Conservation* 22 (1977), pp. 85–102.

Thomson, G. "Air Pollution: A Review for Conservation Chemists." *Studies in Conservation* 10 (1965), pp. 147–167.

Toishi, K., and M. Koyano. "Conservation and Exhibition of Cultural Property in the Natural Environment in the Far East." pp. 90–94 in *The Conservation of Far Eastern Art* (edited by J.S. Mills, P. Smith, and K. Yamasaki). London: International Institute for Conservation of Historic and Artistic Works, 1988.

Uchiyama, S., and S. Hasegawa. "A Reactive and Sensitive Diffusion Sampler for the Determination of Aldehydes and Ketones in Ambient Air." *Atmospheric Environment* 33 (1999), pp. 1999–2005.

Unger, M., M. Stratmann, and R.E. Lobnig. "The Influence of the Amount of Ammonium Sulfate Particles on the Atmospheric Corrosion Mechanism of Copper." pp. 949–960 in *Symposium on the Passivity and its Breakdown, Volume 97-26.* Paris: Electrochemical Society Proceedings, 1998.

Van Bogart, J.W.C. *Magnetic Tape Storage and Handling: A Guide for Libraries and Archives.* Washington, DC: The Commission on Preservation and Access, 1995, p. 28.

Van Bommel, M., B. van Elst, and F. Broekens. "Emission of Organic Acids from Wooden Construction Materials in a Small Test Chamber; Preliminary Results of Optimisation of the Solid Phase Micro Extraction Technique." In *IAP Copenhagen 2001* (edited by M. Ryhl-Svendsen). Copenhagen: Indoor Air Pollution Working Group, 2001. <iaq.dk/iap/iap2001/2001_15.htm>

Van den Brink, O., G.B. Eijkel, and J.J. Boon. "Dosimetry of Paintings: Chemical Changes in Test Paintings as Tools to Assess the Environmental Stress in the Museum Environment." pp. 70–76 in *Preprints of Site Effects: The Impact of Location on Conservation Treatments.* Edinburgh: Scottish Society for Conservation and Restoration, 1998.

Verein Deutscher Ingenieure. *Hygienic Standards for Ventilation and Air-conditioning Systems Offices and Assembly Rooms, VDI 6022.* Dusseldorf: Verein Deutscher Ingenieure, 1998.

Volent, P., and N.S. Baer. "Volatile Amines as Corrosion Inhibitors in Museum Humidification Systems." *International Journal of Museum Management and Curatorship* 4 (1985), pp. 359–364.

Wahlin, P., F. Palmgren, A. Afshari, L. Gunnarsen, O.J. Nielsen, M. Bilde, and J. Kildeso. "Indoor and Outdoor Particle Measurement in a Street Canyon in Copenhagen." pp. 182–187 in *Proceedings of the 9th International Conference on Indoor Air Quality and Climate, Volume 1* (edited by H. Levin). Santa Cruz: International Conference on Indoor Air Quality and Climate, 2002.

Waller, C. *Sorbents for Gaseous Pollutants in Showcases: Anti-tarnish Products for Silver.* 1999. <www.cwaller.de/sorbents.htm>

Waller, R. Canadian Museum of Nature. Personal communication, 2002.

Waller, R. "Internal Pollutants, Risk Assessment and Conservation Priorities." pp. 113–118 in *Preprints of ICOM-CC Conference, Volume 1* (edited by J. Bridgland). London: James & James Science Publishers Ltd., 1999.

Wang, T.C. "A Study of Bioeffluents in a College Classroom." *ASHRAE Transactions* 81 (1975), pp. 32–44.

Watts, S.F. "The Mass Budgets of Carbonyl Sulfide, Carbon Disulfide and Hydrogen Sulfide." *Atmospheric Environment* 34 (2000), pp. 761–779.

Watts, S.F. Communication at the meeting, *Detection and Prevention of Indoor Air Pollution.* The Netherlands Institute for Cultural Heritage, Amsterdam, The Netherlands, August 26, 1999.

Weintraub, S. " Demystifying Silica Gel." In *Objects Specialty Group Postprints, Volume 9* (edited by V. Greene). Washington, DC: Objects Specialty Group of the American Institute for Conservation of Historic and Artistic Works, 2002, pp. 169–194.

Weschler, C.J, H.C. Shields, and D.V. Naik. "Indoor Ozone Exposures." *Journal of the Air Pollution Control Association* 39 (1989), pp. 1562–1568.

Weyde, E. "A Simple Test to Identify Gases which Destroy Silver Images." *Photographic Science and Engineering* 15 (1972), pp. 283–286.

White, K., S. Smith, and P. Dingle. "HEPA Air Filtration: An Effective Method of Reducing Household PM Exposure." pp. 891–895 in *Proceedings of the 9th International Conference on Indoor Air Quality and Climate. Volume 1* (edited by H. Levin). Santa Cruz: International Conference on Indoor Air Quality and Climate, 2002.

Whitmore, P.M., and G.R. Cass. "The Fading of Artists' Colorants by Exposure to Atmospheric Nitrogen Dioxide." *Studies in Conservation* 34 (1989), pp. 85–97.

Whitmore, P.M., and G.R. Cass. "The Ozone Fading of Traditional Japanese Colorants." *Studies in Conservation* 33 (1988), pp. 29–40.

Whitmore, P.M, G.R. Cass, and J.R. Druzik. "The Ozone Fading of Traditional Natural Organic Colorants on Paper." *Journal of AIC* 26 (1987), pp. 45–58.

Wilhelm, H. *The Permanence and Care of Color Photographs: Traditional and Digital Color Prints, Color Negatives, Slides, and Motion Pictures.* Grinnell: Preservation Publishing Company, 1993, pp. 565–568.

Wilke, O., O. Jann, and D. Brodner. "VOC- and SVOC- Emissions from Adhesives, Floor Coverings and Complete Floor Structures." pp. 962–967 in *Proceedings of the 9th International Conference on Indoor Air Quality and Climate, Volume 1* (edited by H. Levin). Santa Cruz: International Conference on Indoor Air Quality and Climate, 2002.

Williams, E.L., E. Grosjean, and D. Grosjean. "Exposure of Artists' Colorants to Peroxyacetyl Nitrate." *Journal of AIC* 32 (1993a), pp. 59–79.

Williams, E.L., E. Grosjean, and D. Grosjean. "Exposure of Artists' Colorants to Sulfur Dioxide." *Journal of AIC* 32 (1993b), pp. 291–310.

Williams, E.L., E. Grosjean, and D. Grosjean. "Exposure of Artists' Colorants to Airborne Formaldehyde." *Studies in Conservation* 37 (1992), pp. 201–210.

Williams, E.L., E. Grosjean, and D. Grosjean. *Exposure of Deacidified Paper to Ambient Levels of SO_2 and NO_2.* GCI Scientific Program Report. Marina del Rey: The Getty Conservation Institute, 1990.

Williams, R.S. "Amines in Steam Humidification Systems." *Conservation DistList.* 2001. <palimpsest.stanford.edu/byform/mailing-lists/cdl/2001/0190.html>

Williams, R.S. "Blooms, Blushes, Transferred Images and Mouldy Surfaces: What are these Distracting Accretions on Art Works?" pp. 65–84 in *Proceedings of the 14th Annual IIC-CG Conference* (edited by J.G. Wellheiser). Ottawa: IIC - Canadian Group, 1989.

Wilson, W.K. *Environmental Guidelines for the Storage of Paper Records. NISO-TR01-1995.* Bethesda, MD: National Information Standards Organization, 1995, p. 8.

WRI (World Resources Institute). *Acid Rain: Downpour in Asia?* nd. <www.wri.org/wri/trends/acidrain.html>

Yamada, E., M. Kimura, K. Tomozawa, and Y. Fuse. "Simple Analysis of Atmospheric NO_2, SO_2 and O_3 in Mountain by using Passive Samplers." *Environmental Science & Technology* 33 (1999), pp. 4141–4145.

Yoon, Y.H., and P. Brimblecombe. "Contribution of Dust at Floor Level to Particle Deposit within The Sainsbury Centre for Visual Arts." *Studies in Conservation* 45 (2000), pp. 127–137.

Yu, D., S.A. Klein, and D.T. Reindl. "An Evaluation of Silica Gel for Humidity Control in Display Cases." *WAAC Newsletter* 23 (2001), pp. 14–19.

Zinn, E., J.M. Reilly, P.Z. Adelstein, and D.W. Nishimura. "Air Pollution Effects on Library Microforms." pp. 195–201 in *Preventive Conservation: Practice, Theory and Research* (edited by A. Roy and P. Smith). London: International Institute for Conservation of Historic and Artistic Works, 1994.

INDEX

Absorption, *see* Sorption

Accelerated ageing 25, 28
    definition 139

Acetic acid 7, 10, 19, 22, 25, 31, 38, 39, 49, 65, 79
    coatings and adhesives as sources of 8
    damage caused by 2, 9, 26, 62, 106
        cellulose acetate 10
        lead 23, 26, 28, 58
        zinc 23, 26
    performance targets 33, 68
    sorption 56
    tests for presence of 82, 83, 84
    various sources of 8, 32, 56, 99
    wood products as sources of 8, 10

Acid deposition *or* acid rain 12, 13, 15

Activated alumina 46
    impregnated with potassium permanganate 45
    see also Filters *and* Sorbents

Activated carbon 45, 46, 56
    activated charcoal cloth 45, 52, 57
    see also Filters *and* Sorbents

Adsorption, *see* Sorption

Adhesives
    acetic acid emission 8, 10, 38, 39, 99
    as sources of pollutants 17, 49, 50, 61, 101

Adverse effect 21, 22, 65, 66, 67, 68, 75
    criteria 106
    definition 139

Aerodynamic diameter 14, 15, 43
    definition 139

Aerosol 8, 12
    definition 139

Ageless 59
    see also Anoxic environment

Agents of deterioration 25, 48, 68, 70, 75, 76, 79, 80, 85
    definition 139

Air cleaners 14, 47

Air-conditioning 14, 41, 44, 87
    see also HVAC system

Air exchange rate 48, 49, 50, 51, 54, 55, 58, 59, 61, 74
    definition 139
    leaky cases or enclosures 50, 70
    measurement 43, 137

Air quality indices 78, 79

Airborne pollutants 7, 8, 9, 17, 21, 28, 31, 32
    definition 139, 142
    fluctuation 79
    test for presence of 82, 83, 84
    see also Key airborne pollutants

Aldehydes 8, 9, 18, 19
    see also Formaldehyde

Alkaline 46, 52

Aluminum foil, barrier 53, 54

Ambient air, definition 139

Amines 8, 9, 85, 100, 108
    diethylamino ethanol (DEAE), corrosion 108
    inhibitor in humidification systems 69, 100

Ammonia (NH_3) 12, 18, 79
    ammonium nitrate 18
    ammonium sulphate 100, 108
    damage caused by 9, 108
    sources of 8, 32, 40, 100
    tests for presence of 83

Analytical hierarchy process (AHP) 70

Anoxic environment 59, 60, 108
    definition 139

Anthropogenic sources 10, 12
    definition 139

ASHRAE 41, 42, 43, 47, 87
    definition 42
    standards 44

Balance (effort, targets) 25, 68, 76

Barrier 29, 35, 59
    airtightness 49
    aluminium foil 53, 54
    block on building level 42
    block on enclosure level 50
    coatings (paints and varnishes), sealing 52, 53, 80
    permeability coefficient 135
    plastic films or bags 52, 60
    protective film on objects 29, 36
    see also Air exchange rate

Bronze 9, 10, 22